essentials for living

a new approach to junior certificate home economics

eilis flood

my-etest

Test yourself on our FREE website www.my-etest.com
Check out how well you score!
Just register once, keep your email as your password, and come back to test yourself regularly.
Packed full of extra questions, my-etest.com lets you revise – at your own pace – when you want – where you want.

GILL & MACMILLAN

Gill & Macmillan Ltd
Hume Avenue
Park West
Dublin 12
with associated companies throughout the world
www.gillmacmillan.ie

© Eilis Flood 2003
© Artwork Vanessa Soodeen 2003; Moira MacNamara 2003
ISBN-13: 978 07171 3473 1
ISBN-10: 0 7171 3473 3
Design by Anú Design, Tara, Co. Meath
Print origination in Ireland by
Compuscript Ltd., Shannon, Co. Clare
Colour reproduction by Ultragraphics, Dublin

The paper used in this book is made from the wood pulp of managed forests.
For every tree felled, at least one tree is planted, thereby renewing natural resources.

All rights reserved. No part of this publication may be copied, reproduced or transmitted in any form or by any means without written permission of the publishers or else under the terms of any licence permitting limited copying issued by the Irish Copyright Licensing Agency.

Contents

Unit 1 Food

1.	Nutrition	2
2.	Digestion	18
3.	Planning the diet	21
4.	Special diets	26
5.	Meal planning	33
6.	Good food hygiene and storage	38
7.	Preparing to cook	45
8.	Cooking food	52
9.	The practical cookery exam	58
10.	The food groups	66
11.	Breakfasts, packed meals, soups and sauces	98
12.	Food processing, food preservation and convenience foods	109
13.	Home baking	116
14.	Recipes	123

Unit 2 Consumer studies

15.	What is a consumer?	154
16.	Consumer protection	157
17.	Making a complaint	161
18.	Quality	164
19.	Money management	169
20.	Shopping	175
21.	Advertising	180

Unit 3 Social studies

22.	The family and adolescence	186
23.	The human body	196
24.	Health education	203

Unit 4 Resource management and home studies

25.	Home management	220
26.	Home design	223
27.	Room planning	228
28.	Services to the home	230
29.	Technology in the home	240
30.	Home hygiene	247
31.	Safety and first aid	249
32.	Community services and the environment	255

Unit 5 Textile studies

33.	Textiles	262
34.	Fashion and design	266
35.	Fibres and fabrics	270
36.	Textile skills	280
37.	Fabric care	290
38.	Practical work	296

Unit 6 Options

39.	Childcare option	300
40.	Design and craftwork option	319
41.	Textile skills option	322
	Picture credits	330

Acknowledgements

I would really like to express my sincere thanks to all who assisted in the production of this book. In particular I would like to thank my husband John for his continued support, and my parents, sisters and brother for their tremendous interest and encouragement. I would like to thank everyone at Gill & Macmillan, especially Hubert Mahony, Tess Tattersall and Anna Scobie, and also Julia Fairlie, Vanessa Soodeen and Moira MacNamara.

Eilis Flood
March 2003

Unit 1

Food

chapter 1

Nutrition

Why do we need food?

Malnutrition – what is it?

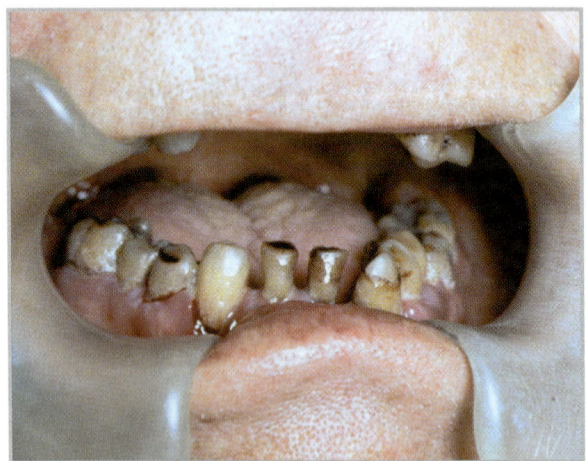

Too little of certain foods, for example fruit and vegetables, causes scurvy

Too much of certain foods

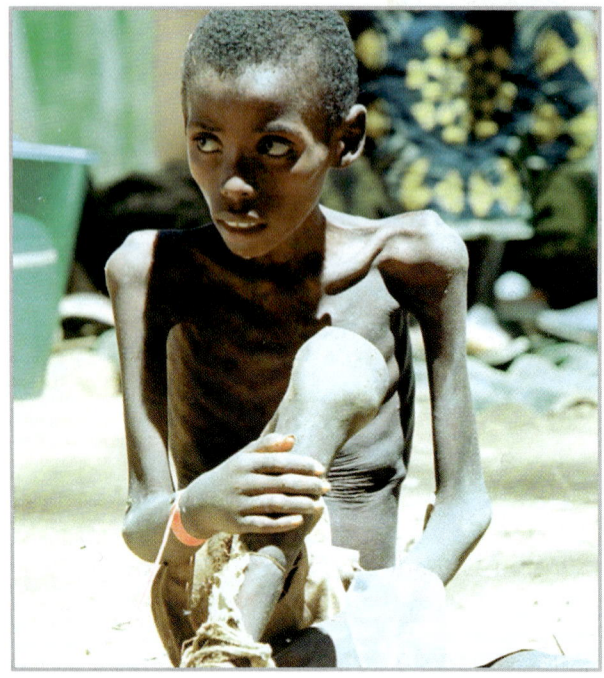

Too little food

Nutrients

All food contains one or more of the six nutrients. The names of the nutrients are:
- protein
- fat
- carbohydrate
- minerals
- vitamins
- water

Each and every one of the six nutrients must be taken in to keep our bodies healthy.

Protein, fat and carbohydrate are eaten in large amounts. They must be broken down by the body during digestion and are called *macronutrients*.

Minerals and vitamins are needed in smaller amounts and need no digestion. They are called *micronutrients*.

PROTEIN

Composition of protein

Protein is made up of chains of amino acids. Each amino acid is made up of the elements carbon, hydrogen, oxygen and nitrogen (CHON). Of these, nitrogen is the most important one as it is needed for growth of the body. During digestion enzymes break the protein chains down into single amino acids. When protein reaches this simple form the body is able to use it.

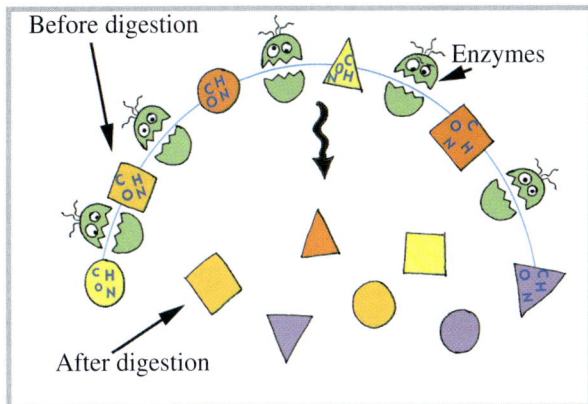

Classification of protein

First class protein

Animal protein, e.g. meat, fish, eggs, cheese.

Second class protein

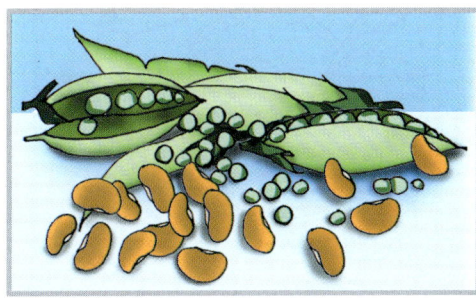

Vegetable protein, e.g. peas, beans, pasta, nuts, brown bread.

ESSENTIALS FOR LIVING

Essential amino acids

Higher level

You learned earlier that protein is made up of chains of amino acids. While all amino acids are good for the body, some are more important than others. The most important are called *essential amino acids*.

High and low biological value protein

Some protein foods contain most or all of the essential amino acids and are called *first class* or *high biological value* (HBV) protein foods. These foods usually come from animal sources such as meat. (The main exception is TVP, which is vegetable protein with a high biological value.) Other protein foods do not contain all the essential amino acids and are called *second class* or *low biological value* (LBV) protein foods. These are usually from vegetable or plant sources such as beans.

Sources: What foods contain protein?

Sources of animal protein

Red meat, fish, eggs, cheese, yogurt, milk, chicken.

Sources of vegetable protein

Peas, beans, lentils, pasta, rice, potatoes, nuts, brown bread, breakfast cereals.

 Activity 1.1 – Workbook p. 3

 Activity 1.2 – Workbook p. 4

Functions: What does protein do in the body?

- Allows growth of the body
- Repairs worn out and damaged cells e.g. when you cut yourself
- Makes enzymes, hormones and antibodies
- Gives us heat and energy

NUTRITION

Recommended daily allowance (RDA)

Higher level

This tells us how much of each of the nutrients we need per day. The table below shows the RDA for some of the nutrients needed by the body.

Nutrient	RDA
Protein	1 g / kg of body weight
Fat	no more than 30% of energy intake
Carbohydrate	depends on how active the person is
Fibre	30 g
Vitamin C	30 mg
Vitamin D	2.5–10 mg
Vitamin A	750–1,200 mg
Iron	10–15 mg
Calcium	500–1,200 mg
Sodium (salt)	2 g

 Activity 1.3 – Workbook p. 4

Textured vegetable protein (TVP)

Textured vegetable protein or TVP is a meat substitute generally made from soya beans. TVP is an excellent food in that it is high in protein and calcium, yet low in fat. It is available in steaks, chunks or minced form.

A meal made with TVP

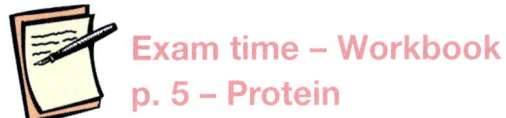 **Exam time – Workbook p. 5 – Protein**

FATS

Composition of fats

Fats are made up of fatty acids and glycerol. Each glycerol molecule has three fatty acid molecules attached to it. During digestion enzymes break fat molecules down.

Fats contain the elements carbon, hydrogen and oxygen (CHO).

ESSENTIALS FOR LIVING

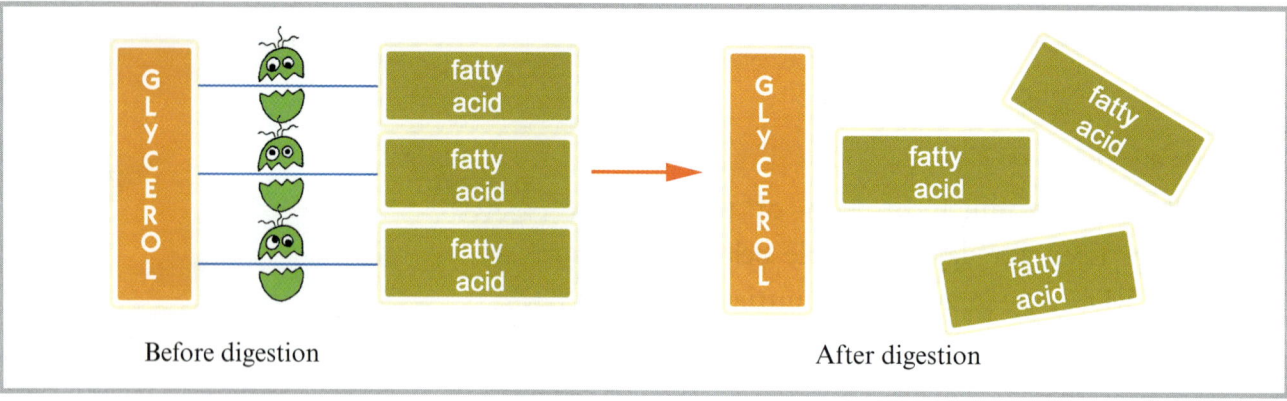

During digestion enzymes break the links between glycerol and fatty acids

Classification of fats

There are two types of fat: saturated and unsaturated.

Saturated fats

These come from animals and are sometimes called animal fats.

Unsaturated fats

These come from fish and plants and are sometimes called vegetable fats.

Sources: What foods contain fat?

Saturated fats

Meat such as rashers, sausages, chops, cream, butter, eggs, milk.

Unsaturated fats

Polyunsaturated margarine such as Flora, beans, nuts, fish, fish liver oils such as cod liver oil, cereals.

 Activity 1.4 – Workbook p. 6

Functions: What does fat do in the body?

NUTRITION

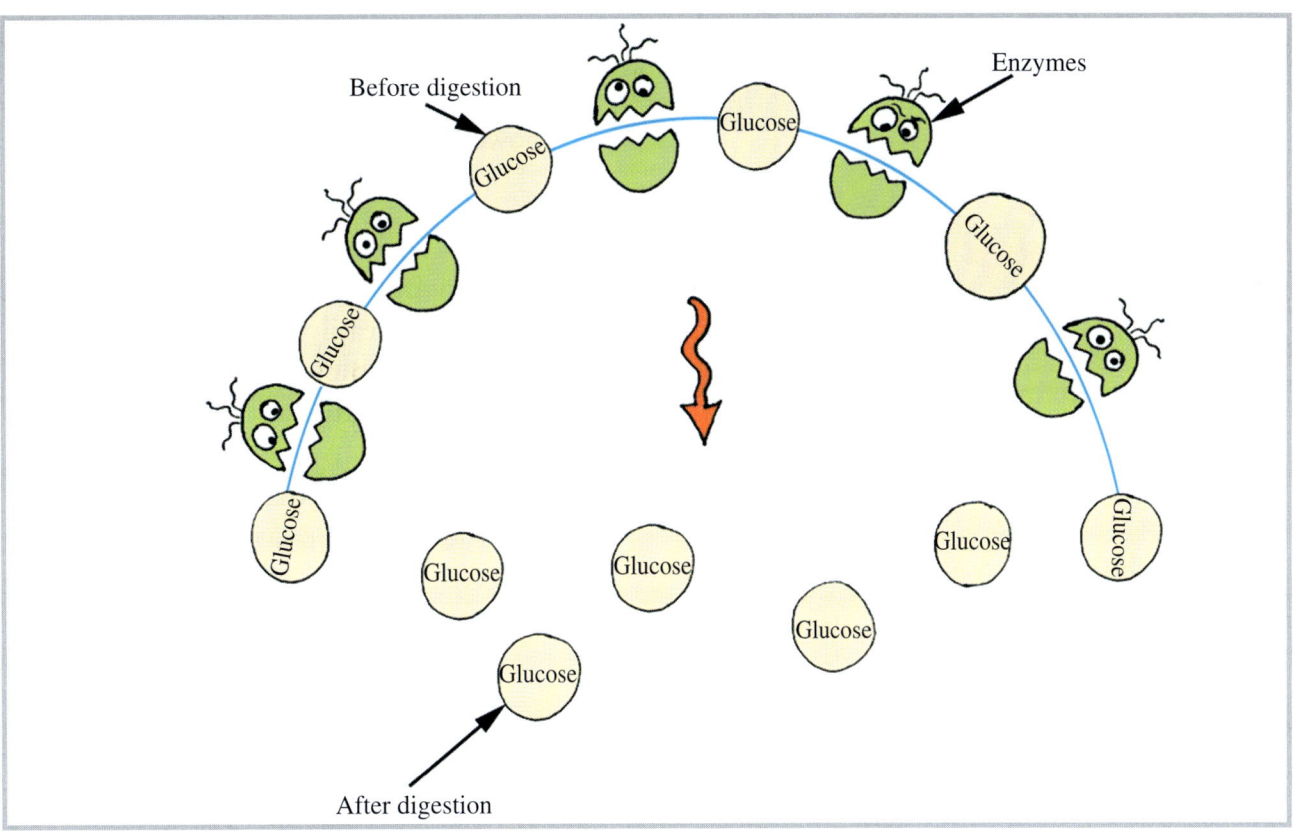

During digestion enzymes break carbohydrates into simple sugars

- Gives us lots of heat and energy
- A layer of fat under the skin called adipose tissue insulates the body (helps keep body warm)
- Delicate organs such as heart and kidney have a layer of fat around them; this layer can protect them from injury
- Vitamins A, D, E and K dissolve in fat; when we eat fat we eat these vitamins also

How to cut down on fat

- Grill, bake, boil or microwave instead of frying
- Use polyunsaturates, e.g. Flora, instead of butter
- Cut fat off meat
- Cut down on foods like crisps, burgers and chips
- Use low-fat foods

Activity 1.5 – Workbook p. 7

Exam time – Workbook p. 7 – Fats

CARBOHYDRATES

Composition of carbohydrates

Like fats, carbohydrates are made up of the elements carbon, hydrogen and oxygen (CHO). Fats produce more energy than carbohydrates as they have more carbon. Carbohydrates are made up of chains of simple sugars.

Classification of carbohydrates

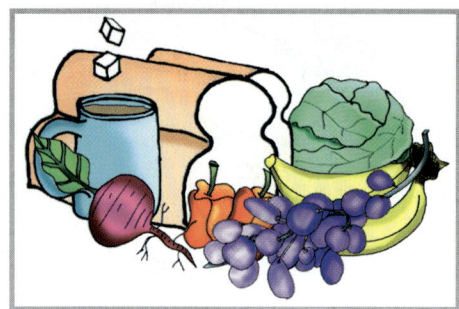

Foods that contain carbohydrate can be divided into three groups:
- sugars
- starches
- fibre-rich foods

Sources: What foods contain carbohydrate?

Sugars: honey, fruit, ice cream, biscuits, cake, chocolate, sweets.

Starches: bread, breakfast cereals, potatoes, pasta, other vegetables such as carrots and broccoli.

Fibre-rich foods: brown bread, brown rice, fruits, vegetables, fibre-rich breakfast cereals.

 Activity 1.6 – Workbook p. 8

 Activity 1.7 – Workbook p. 8

Functions: What does carbohydrate do in the body?

- Provides heat and energy
- Fibre-rich carbohydrates fill you up
- Fibre helps stop constipation and other bowel problems

FIBRE

Fibre is very important in the diet because it helps waste pass easily through the intestines. If we do not eat enough fibre we risk constipation and other diseases such as cancer of the bowel.

It is recommended that we eat 25–30 g of fibre per day, yet on average we eat only 15–20 g.

Refined carbohydrates

Refined or processed foods rarely contain much fibre.

To increase fibre:

- Eat more brown bread and rice instead of white
- Eat more fresh fruit and vegetables
- Eat high-fibre breakfast cereals such as bran flakes

 Activity 1.8 – Workbook p. 9

 Activity 1.9 – Workbook p. 10

EMPTY-CALORIE FOODS

Most of the foods pictured above contain lots of sugar but very little goodness. These foods are all high-calorie foods but do not supply any of the other nutrients needed by the body. This is why they are called 'empty-calorie' foods.

Hidden sugar

The foods pictured above are obviously sugary foods. Sometimes, however, foods have hidden sugar: watch out for the words sucrose, glucose, maltose, fructose and dextrose – these are all types of sugar.

Water-soluble vitamins

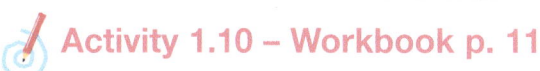
Activity 1.10 – Workbook p. 11

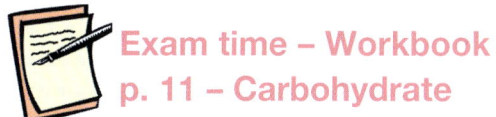
Exam time – Workbook p. 11 – Carbohydrate

Vitamin	Functions	Sources	RDA	Deficiency diseases
C	• general health • healthy skin and gums • helps wounds to heal • helps absorb iron	*Fruit:* oranges, blackcurrants, lemons, grapefruits, peppers, strawberries, orange juice *Vegetables:* cabbage, peppers, tomatoes, new potatoes, coleslaw	• 30 mg for adults; • more for adolescents	• scurvy • wounds slow to heal
B group (6 vitamins)	• growth • healthy nervous system • controls the release of energy from food	nuts, peas, beans, lentils, yeast bread, brown bread, meat, fish, milk, eggs, cheese	varies with each vitamin	• beri-beri (nerve disease): may appear with alcoholism • pellagra: sores on the skin and tongue

ESSENTIALS FOR LIVING

VITAMINS

Vitamins are needed in small amounts by the body. There are two groups of vitamins:

Water-soluble → These dissolve in water: Vitamins C and B group

Fat-soluble → These dissolve in fat: Vitamins A, D, E and K

Fat-soluble vitamins can be stored in the body but water-soluble vitamins cannot and need to be eaten every day.

✎ Activity 1.11 – Workbook p. 12

✎ Activity 1.12 – Workbook p. 12

✎ Activity 1.13 – Workbook p. 13

Fresh fruit and vegetables contain vitamin C which prevents scurvy

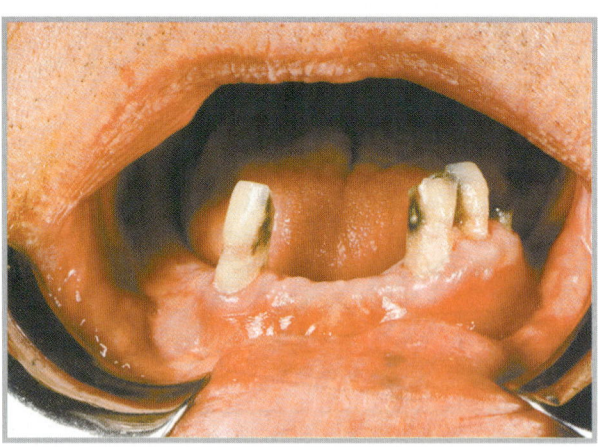

Scurvy

Scurvy is a disease caused by lack of vitamin C. It was once a common disease among sailors who survived mainly on salt beef and hardtack (dry biscuits). These men rarely had fresh fruit and vegetables and frequently died from the disease.

In 1573 a Scottish doctor called James Lind discovered that eating oranges and lemons cured scurvy. The symptoms of scurvy are:
- poor healing of wounds
- frequent bruising
- sore, bleeding mouth and gums
- anaemia

Scurvy, while rare, is still found in infants and elderly people living on poor diets.

Babies, especially bottle-fed babies, should be given fruit juice in a cup or from a spoon from one month old.

Preventing vitamin C loss

Vitamin C is easily destroyed. To help prevent this loss:
- Don't buy wilted vegetables
- Cook vegetables soon after buying
- Cook vegetables in a small amount of water, bring to the boil and then simmer

10

- Don't overcook vegetables

- Don't use bread soda when cooking cabbage as it destroys vitamin C
- Eat vegetables soon after cooking: vitamin C is lost if they are kept warm for a long time

Rickets and osteomalacia

The disease rickets is caused by lack of vitamin D. It was first diagnosed in the 1650s and was common in children living in the industrial smog-filled cities at the beginning of the twentieth century. Rickets, while rare nowadays, does still appear in a small number of infants from low-income families. Some vegetarian children and infants who have been

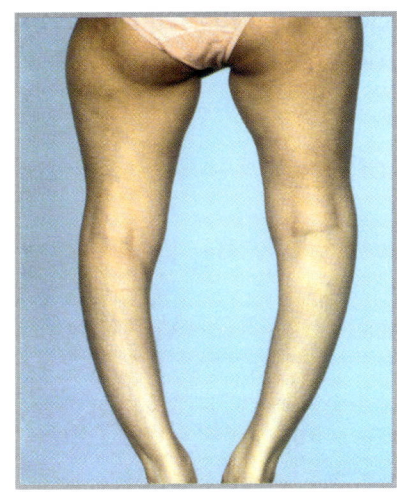

breast-fed for a long time without supplements are also at risk. Because vitamin D is needed to absorb or use calcium, and calcium is needed to harden the bones, a child with rickets will have soft, badly formed bones (bow legs) and teeth that are prone to decay.

Osteomalacia

Osteomalacia is the adult form of rickets. In developing countries this disease is found in women who have low intakes of vitamin D and calcium. Many of these women have had

Fat-soluble vitamins

Vitamin	Sources	Functions	Deficiency symptoms
A	*Pure vitamin A:* fish liver oils, oily fish, kidney, liver, margarine, eggs, milk *Carotene:* highly coloured vegetables e.g. carrots	• growth • healthy eyes • healthy skin • healthy linings of the nose and throat etc.	• slowed growth • night blindness • linings of the nose and throat become dry and irritated
D	sunshine, fish liver oils, oily fish, margarine, liver, cheese, eggs	• works with calcium for healthy bones and teeth	• rickets (children) • osteoporosis (adults)
K	made in the bowel, green vegetables, cereals	• helps the blood to clot	• problems with blood clotting

several closely spaced pregnancies followed by long periods of breast feeding.

 Activity 1.14 – Workbook p. 13

 Activity 1.15 – Workbook p. 14

 Activity 1.16 – Workbook p. 14

 Activity 1.17 – Workbook p. 15

 Exam time – Workbook p. 15 – Vitamins

MINERALS

Minerals, like vitamins, are needed in small amounts by the body for growth and general good health.

Important minerals include:
- calcium
- iron
- iodine
- fluorine
- sodium
- phosphorus

Calcium

Sources: What foods contain calcium?

Milk, cheese, yoghurt, tinned fish, water, cereals, beans and bread.

Functions of calcium

Calcium is needed to build strong bones and teeth.

Deficiency symptoms and diseases

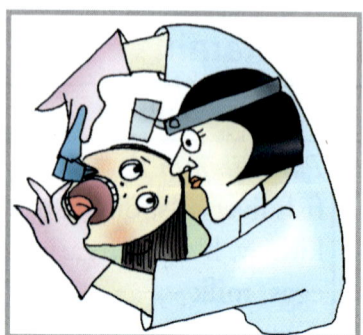

If you don't have enough calcium you risk rickets, osteomalacia (adult rickets), osteoporosis (brittle bones) and tooth decay.

 Activity 1.18 – Workbook p. 16

Iron

Sources: What foods contain iron?
Liver, kidneys, red meat, cabbage, broccoli, green peppers, Brussels sprouts, peas, curly kale, cauliflower, lettuce, spinach, scallions, tomatoes.

 Activity 1.19 – Workbook p. 16

Functions and deficiency of iron
There is a substance called haemoglobin in our blood. Haemoglobin carries oxygen to all our body cells. Iron is needed to make haemoglobin. If there is not enough iron in our diet anaemia is the result. The symptoms of anaemia are tiredness, paleness, weakness and dizziness. (A recent Irish National Nutrition Survey found that approximately seventy per cent of Irish teenage girls were not eating enough iron.)

Salt
Note: Too much *salt* (sodium) is linked with high blood pressure and strokes.

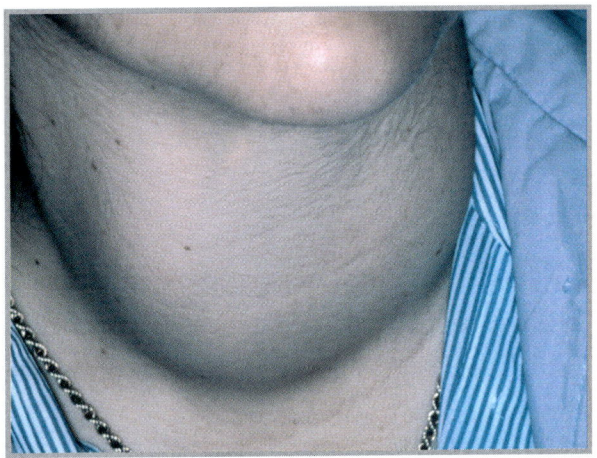

Goitre: a disease of the thyroid gland

Other minerals

Mineral	Sources	Functions	Deficiency symptoms
Iodine	sea fish, seaweed, sea salt, vegetables, cereals, milk	needed for a healthy thyroid gland (in the neck)	goitre (see photo)
Fluorine	sea fish, tap water, toothpaste	helps prevent tooth decay	tooth decay
Sodium (salt)	salt added to food, bacon, snack and convenience foods	needed for water balance in the body	cramps, bloating
Phosphorus	some found in most foods especially high-protein foods	works with calcium for strong bones and teeth	none known

ESSENTIALS FOR LIVING

Vitamin and mineral link

Vitamin C is needed to absorb iron.
Vitamin D is needed to absorb calcium.

 Activity 1.20 – Workbook p. 17

 Activity 1.21 – Workbook p. 17

 Exam time – Workbook p. 17 – Minerals

WATER

Sources of water

Food

All foods contain water except solid fats such as lard, and dried foods such as cornflakes. Fruit and vegetables can consist of up to ninety per cent water.

Drinks

Water, milk, tea, coffee, soft drinks, fruit juice.

Functions: What is water needed for in the body?

- Body fluids, such as blood, are mainly water
- Water is often a source of calcium and fluoride
- Water removes waste from the body, for example in urine
- Water helps keep the body at the right temperature; for instance, we sweat water when we are too hot

Dehydration

Dehydration is when the body loses large amounts of water. It is particularly dangerous in babies. Dehydration can occur after a long period of vomiting or diarrhoea.

 Activity 1.22 – Workbook p. 18

Properties of water

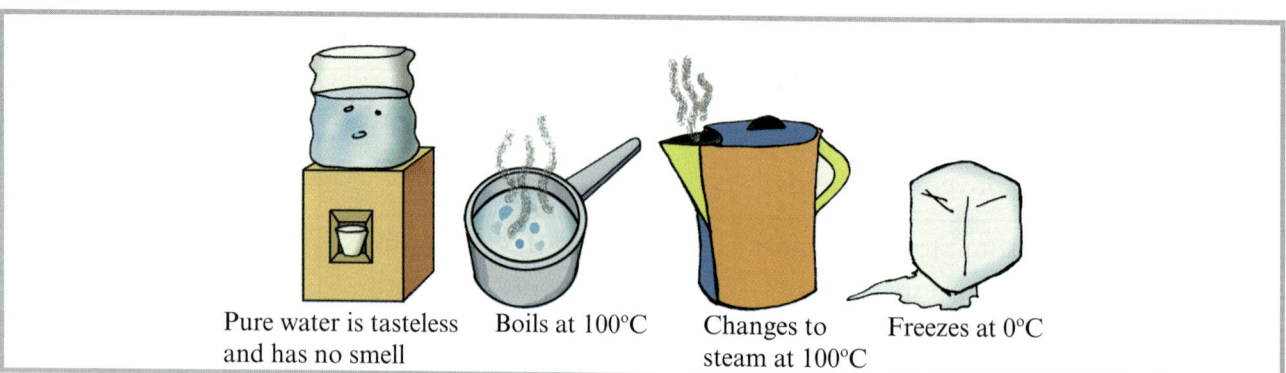

Pure water is tasteless and has no smell   Boils at 100°C   Changes to steam at 100°C   Freezes at 0°C

ENERGY

How energy is produced in the body

Food is burned in the body cells and energy is produced. Oxygen is needed. This process is called *oxidation*.

Energy balance

Energy balance, as the name suggests, is all about energy input (food) equalling energy output (BMR + activity).

Note: BMR is explained on page 17.

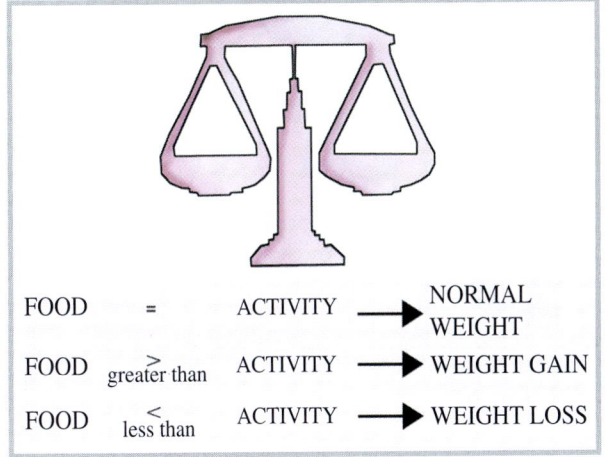

Kilocalories and kilojoules

Just as the amount of milk in a carton is measured in litres, the amount of energy provided by food is measured in kilocalories (kcal) or kilojoules (kJ). (1 kcal = 4.2 kJ)

 Activity 1.23 – Workbook p. 19

Food labels

All foods that contain protein, fat or carbohydrate produce energy and are said to contain calories. Food products must display nutritional information on the package. One piece of information displayed is the food's energy value. Food labels give energy values per 100 g of the food. Food labels often give values for a serving of the food as well.

 Activity 1.24 – Workbook p. 19

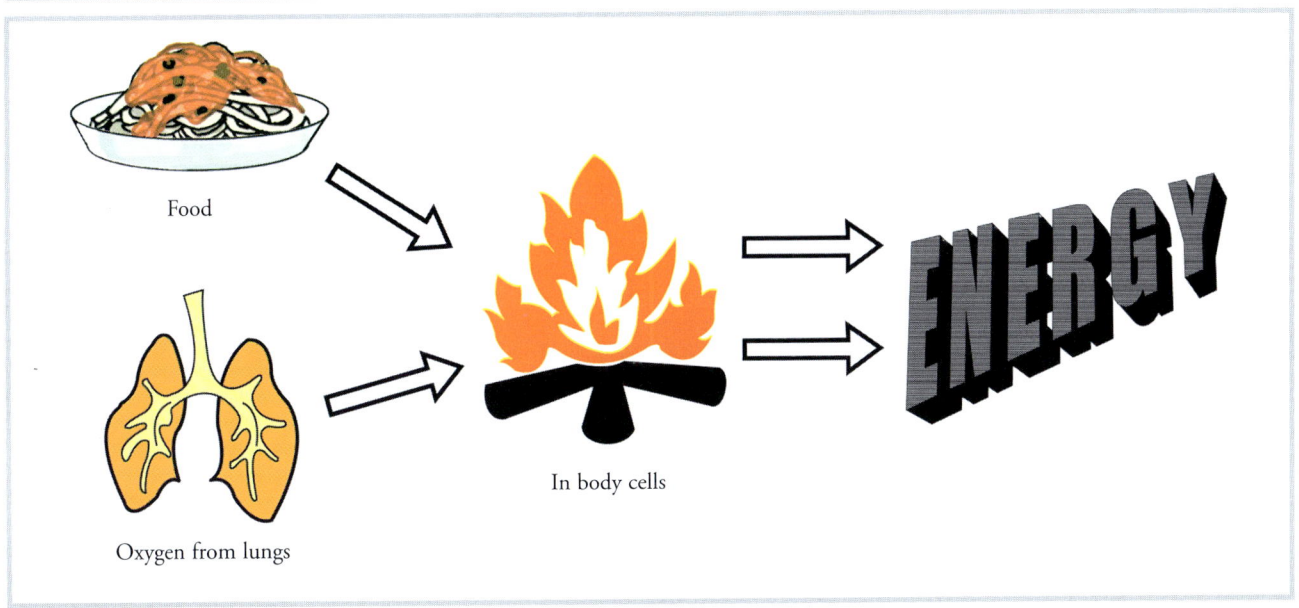

Oxidation

ESSENTIALS FOR LIVING

How much energy do I need?

The amount of energy (number of kilocalories or kilojoules) needed by an individual depends on their size, whether they are male or female, their age and how active they are. Pregnant or breast-feeding women and people living in cold climates need extra energy. However, there are *average* recommended daily allowances (RDAs) for energy.

Average RDA for energy

Approximate daily energy requirements (in kilocalories)

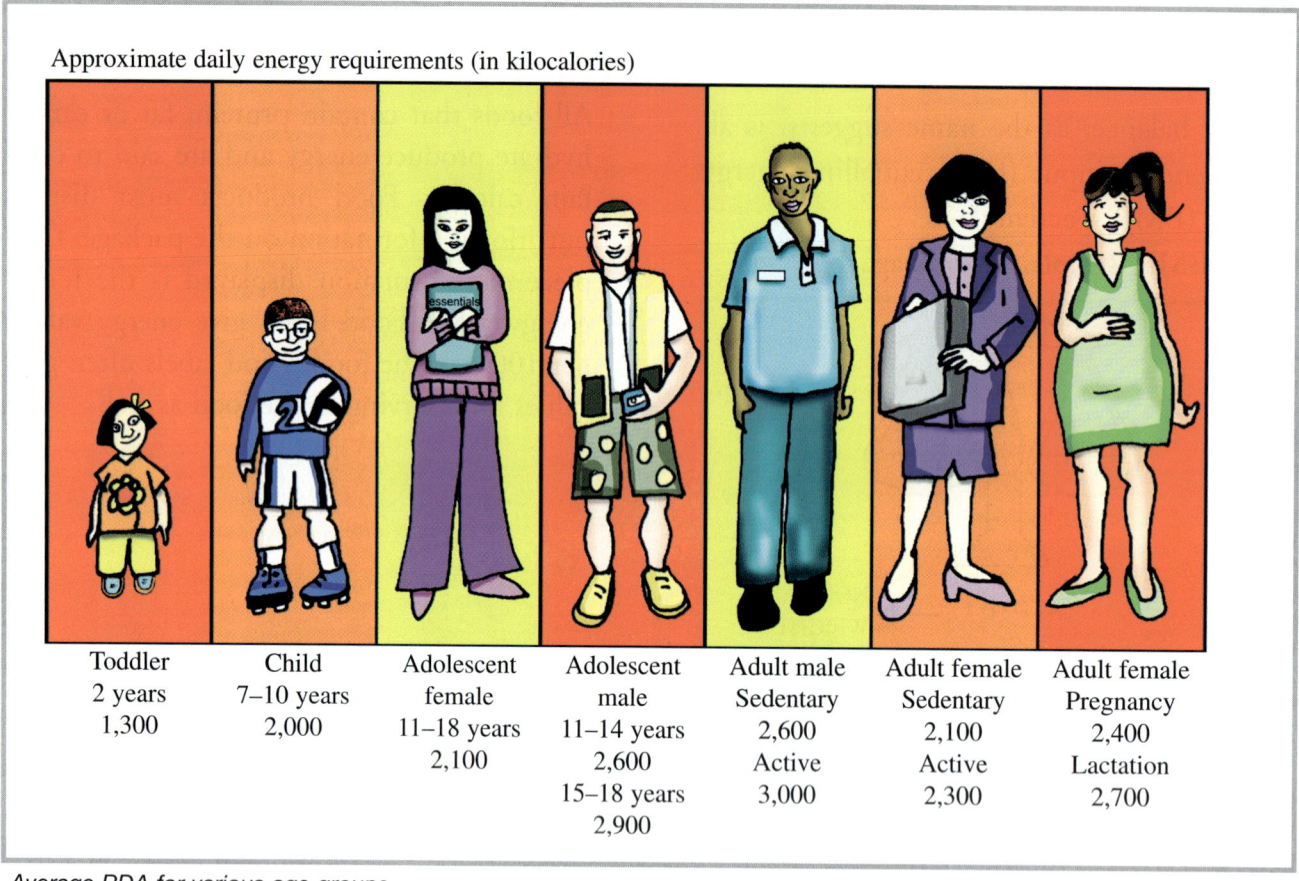

Toddler 2 years	Child 7–10 years	Adolescent female 11–18 years	Adolescent male 11–14 years	Adult male Sedentary	Adult female Sedentary	Adult female Pregnancy
1,300	2,000	2,100	2,600	2,600	2,100	2,400
			15–18 years 2,900	Active 3,000	Active 2,300	Lactation 2,700

Average RDA for various age groups

Energy output

The body needs a basic amount of energy to stay alive; this is called a person's *Basal Metabolic Rate* (BMR). Extra energy is needed for activities. Look at the graph below to see the amount of energy (kilocalories) needed for various activities.

 Activity 1.25 – Workbook p. 20

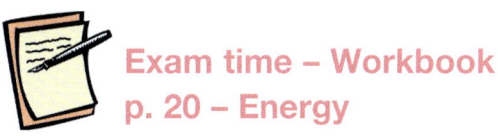 **Exam time – Workbook p. 20 – Energy**

Revision crossword – Workbook p. 21

Now test yourself at *www.my-etest.com*

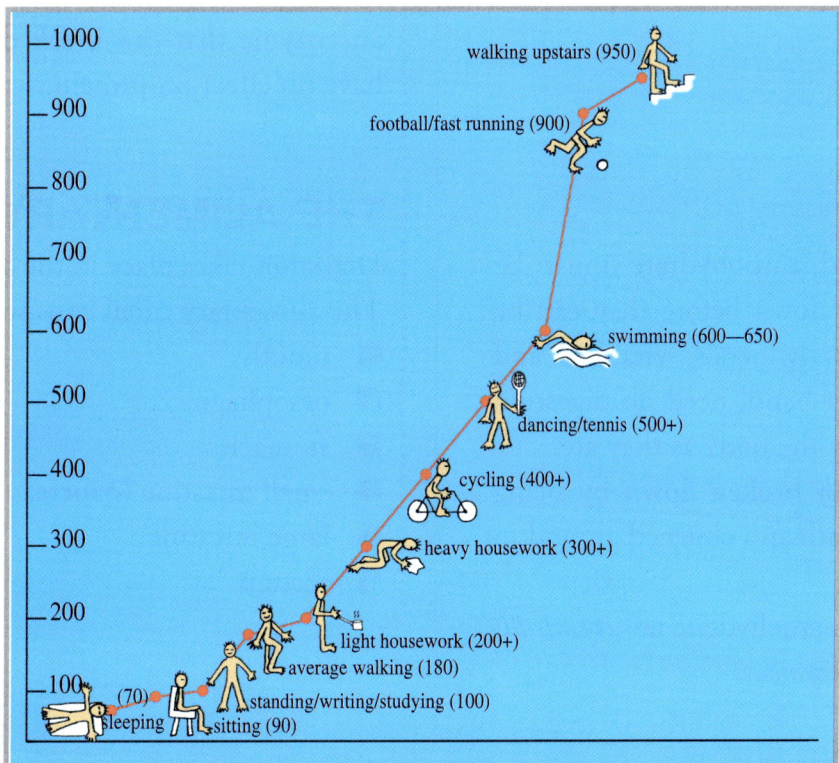

chapter 2

Digestion

Food, as we saw in chapter one, contains the six nutrients:
- protein
- fat
- carbohydrate
- minerals
- vitamins
- water

Protein, fat and carbohydrate must be *digested* or broken down before they can be used by the body. Minerals, vitamins and water, on the other hand, need no digestion and can be used by the body as they are.

Food is *physically* broken down by being chewed in the mouth and churned around in the stomach.

Protein, fat and carbohydrate are *chemically* broken down by *enzymes*.

Digestive enzymes are chemical substances that break down food. Enzymes themselves do not change in the process. Each enzyme can only work on one nutrient, for example an enzyme that can break down fats would have no effect on protein.

THE ALIMENTARY CANAL

Digestion takes place in the alimentary canal. The alimentary canal consists of:
- mouth
- oesophagus
- stomach
- small intestine (pancreas and liver)
- large intestine
- rectum

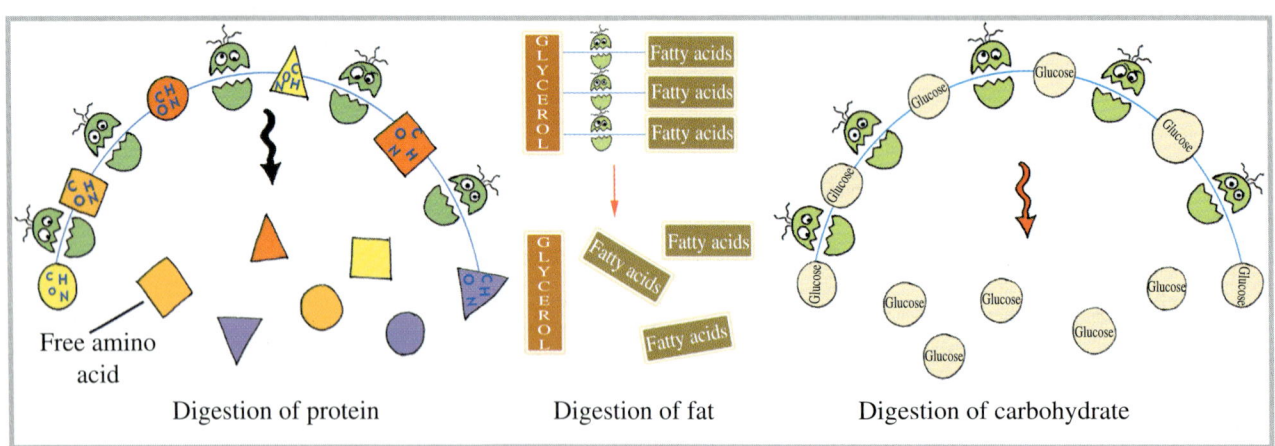

Digestion of protein   Digestion of fat   Digestion of carbohydrate

DIGESTION

1. In the mouth the teeth physically break food down. The tongue mixes it with saliva from the *salivary glands*. Saliva contains the enzyme *amylase*. Amylase breaks down long chains of cooked starch such as bread into shorter chains called *maltose*.

2. Food is then swallowed and passes into a long tube called the *oesophagus*. The walls of the oesophagus move in and out squeezing the food along; this is called *peristalsis*.

3. Food then passes into the *stomach*. The stomach can be described as a muscular bag, which lies in the left-hand side of the abdomen under the *diaphragm*. Food is churned about in the stomach until it turns into a liquid called *chyme*. The walls of the stomach produce a juice called *gastric juice*. This juice is made of a strong acid, *hydrochloric acid*, and two enzymes. One of these enzymes is called *pepsin* and it starts to digest protein by breaking the long amino acid chains into shorter ones called *peptides*.

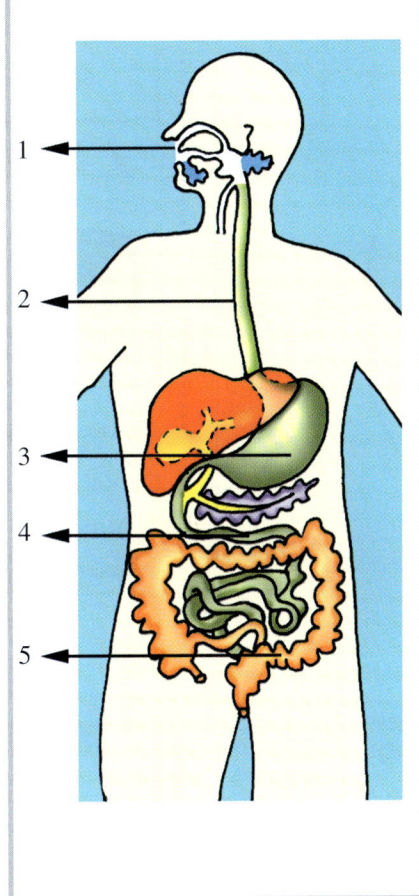

4. Next food passes into the narrow, but very long, *small intestine*. Again peristalsis keeps the food moving along. Three digestive juices work here to complete the digestion of our food:
- pancreatic juice (made by the pancreas)
- bile (made by the liver; works on fats)
- intestinal juice (made by the intestine itself)

All three juices work together to change:
- protein and peptides to free amino acids
- carbohydrates to monomers (simple sugars)
- fats to fatty acids and glycerol

Digested nutrients are then absorbed into the blood. They travel to our body cells where they carry out their various functions, for example producing heat and energy.

5. Waste that is not absorbed now passes into the *large intestine* (*colon* or *bowel*). The waste, as it passes into the large intestine, is still very liquid. Some of this water passes back through the walls of the intestine into the bloodstream. This causes the waste to become more solid. It is now called *faeces*. Faeces passes out of the body through the rectum and anus.

ESSENTIALS FOR LIVING

 Activity 2.1 – Workbook p. 22

The small intestine

The walls of the small intestine resemble a very fine carpet: they are covered in a huge number of hair-like projections called *villi*. Each villus has a tiny blood vessel and a lymph vessel. It is into these vessels that digested nutrients pass.

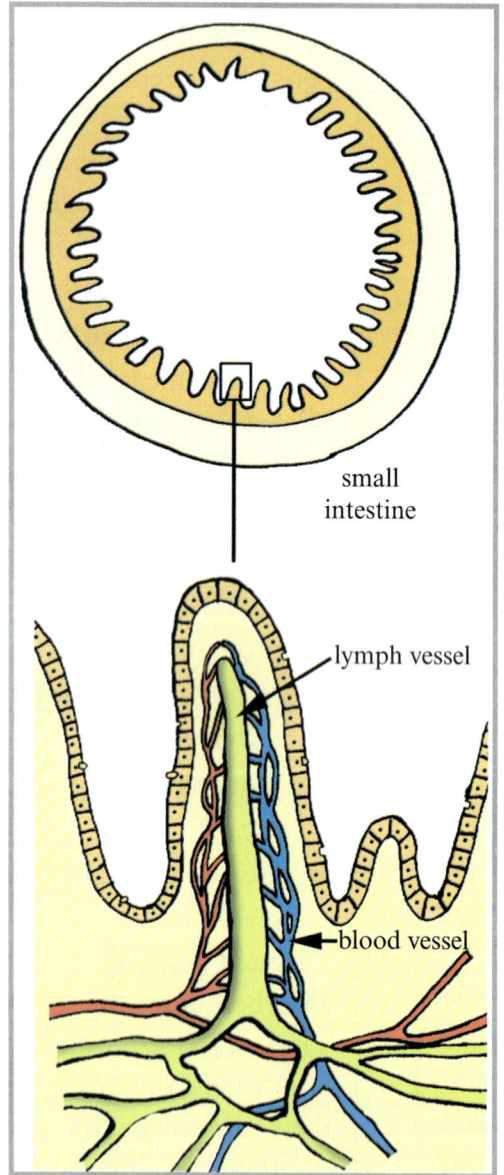

The absorption and transport of protein and carbohydrate is simple:

small intestine → blood vessel of villi → bloodstream → body cells

Fats are more complicated:

small intestine → lymph vessel of villi → lymph system → bloodstream → body cells

 Activity 2.2 – Workbook p. 22

 **Exam time – Workbook p. 23**

Revision crossword – Workbook p. 23

Now test yourself at *www.my-etest.com*

chapter 3

Planning the diet

To have a balanced diet we need to eat the correct amount of each of the six nutrients.

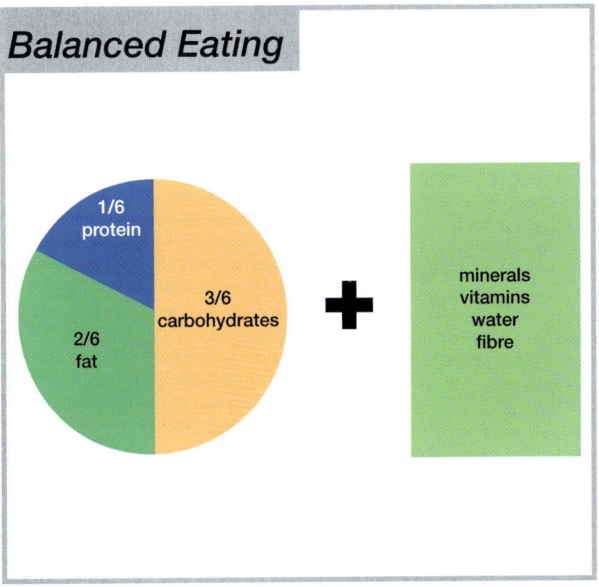

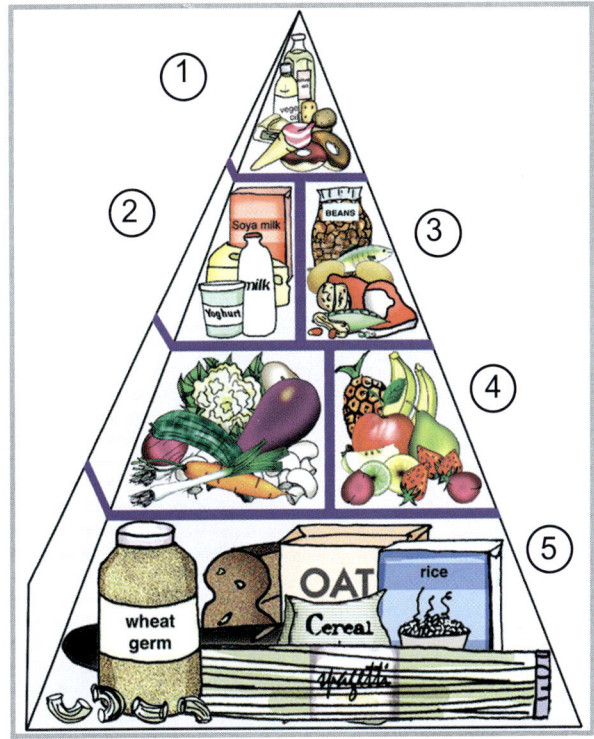

Healthy eating is very important for us to look and feel well. No face creams or health supplements can take the place of eating a wide variety of fresh foods and drinking plenty of water.

Generally we all have some idea of what we should and should not be eating. For many of us, though, working out a balanced diet seems too much like hard work!

The food pyramid pictured here is designed to make planning a healthy, balanced diet easy. It divides food into four groups and then advises us on how much of each group we need each day.

1. Fats, oils, cakes etc. – eat as little as possible
2. Milk and milk products group – eat three to four servings per day
3. Protein group – eat two servings per day
4. Fruit and vegetable group – eat four servings per day
5. Cereal and potato group – eat six servings per day

Activity 3.1 – Workbook p. 25

Activity 3.2 – Workbook p. 26

Activity 3.3 – Workbook p. 27

The food pyramid gives us an excellent general guide to planning a balanced diet for

ESSENTIALS FOR LIVING

an adult or a teenager. At various times in our lives, though, we have other particular dietary needs that require special attention. We will now look at the special dietary needs of each of the following groups:

- babies
- children
- adolescents
- adults
- pregnant and breast-feeding women
- elderly people

Babies

1. Breast feed if possible
2. Prepare formula feeds carefully:
   - sterilise completely
   - measure correctly
   - follow instructions exactly
3. Use formula up to one year; do not use skimmed milk for children as it lacks essential fat and vitamins
4. Do not allow babies and toddlers to sleep with a bottle in their mouth; this causes 'bottle rot' and they can choke
5. Babies need vitamin C. Serve juice in a cup or from a spoon – never from a bottle; it rots teeth
6. From six months babies need iron-rich foods such as dark green vegetables and liver
7. Wean onto puréed solids at four to six months
8. Do not add salt or sugar to a baby's food

 Activity 3.4 – Workbook p. 27

Children

1. Have regular meal times
2. Serve small attractive portions
3. Do not fuss about food refusal; it makes it worse
4. Serve healthy snacks such as fruit or popcorn
5. Continue to use whole milk: it contains vitamin A for growth and vitamin D for healthy bones
6. Try to serve healthy packed lunches; avoid junk food such as sweets, crisps etc.
7. Set a good example yourself

PLANNING THE DIET 3

 Activity 3.5 – Workbook p. 28

Adolescents/teenagers

1. The growth spurt during the teenage years means teenagers need plenty of protein, calcium and iron
2. Teenagers, especially girls, need to stay active to avoid weight gain
3. Girls need to eat iron-rich foods such as red meat and green vegetables to avoid anaemia because of periods (menstruation)
4. For healthy skin avoid fried foods, chocolate etc. Eat fruit and vegetables and drink plenty of water
5. Choose healthy snacks such as yogurt and fruit

 Activity 3.6 – Workbook p. 28

Adults

1. Weight gain becomes more common as people enter adulthood. People with sedentary jobs should cut down on what they eat to avoid weight gain

Sedentary worker *Manual worker*

2. Many people (especially women) stop taking regular exercise once they leave school. Adults should try to take some form of exercise at least three times per week
3. Eat plenty of fibre-rich foods such as brown bread, fruit and vegetables: they prevent bowel disease and are low in calories
4. Do not eat too much salt: it can cause high blood pressure and strokes
5. Avoid foods that have too much saturated fat such as butter and sausages: saturated fats contain cholesterol, which is linked to heart disease
6. Women in particular should eat calcium-rich foods such as milk, yogurt and cheese to avoid brittle bones in later life
7. Avoid too much alcohol
8. Because of monthly blood loss all women risk anaemia. To avoid anaemia eat iron-rich foods such as green vegetables.

Note: vitamin C is also needed to absorb iron

ESSENTIALS FOR LIVING

Pregnant and breast-feeding women

1. Pregnant women are *not* eating for two. Eating 'empty-calorie' foods during pregnancy will cause weight gain, which will be hard to lose afterwards
2. Average weight gain during pregnancy is 12 kg (2 stone)
3. Avoid smoking, alcohol, strong tea, coffee or spicy foods. Avoid lightly cooked eggs, unpasteurised cheeses and cook-chill foods because of the risk of food poisoning
4. Eat iron-rich foods such as green vegetables to avoid anaemia.

Note: vitamin C is needed to absorb iron

5. Take vitamin D to help absorb calcium. Vitamin D is found in cod liver oil, oily fish such as mackerel, and is made in the body in the presence of sunshine

Folic Acid

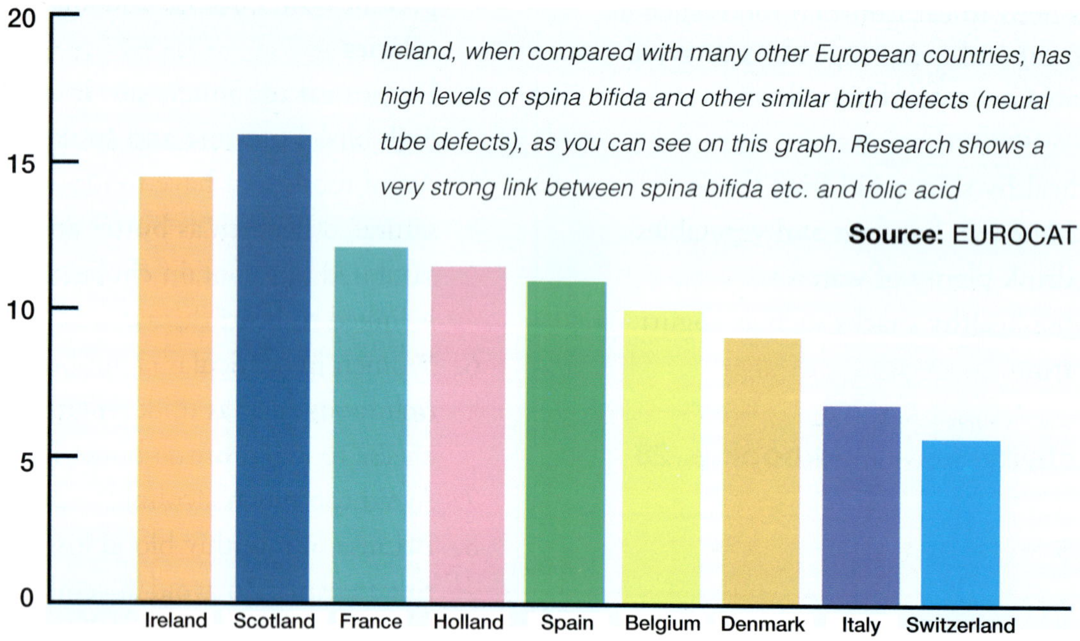

Ireland, when compared with many other European countries, has high levels of spina bifida and other similar birth defects (neural tube defects), as you can see on this graph. Research shows a very strong link between spina bifida etc. and folic acid

Source: EUROCAT

Ireland	14.5
Scotland	16
France	12
Holland	11.5
Spain	11
Belgium	10
Denmark	9
Italy	7
Switzerland	5.5

Folic acid helps prevent spina bifida and should be taken in tablet form for three months before becoming pregnant and for the first twelve weeks of pregnancy. Some foods such as breakfast cereals are fortified with folic acid.

6. Eat calcium for healthy bones and teeth: it is found in milk, tinned fish and cheese (pasteurised only)

 Activity 3.7 – Workbook p. 29

Elderly persons

1. Try to eat three nourishing meals a day
2. Choose low-fat cheese, milk and yogurt: these foods are high in calcium and low in fat. Calcium prevents osteoporosis (brittle bones)
3. To prevent heart disease limit high-cholesterol foods such as butter and eggs: choose polyunsaturates instead, such as Flora
4. Eat plenty of fruit and vegetables; the elderly often lack vitamin C
5. Reduce salt, fat and sugar intake
6. Choose high-fibre foods such as brown bread and fruit; elderly people are prone to constipation

7. Limit intake of processed foods such as tinned foods

 Activity 3.8 – Workbook p. 30

Invalids and convalescents (those recovering from illness)

1. Follow doctor's advice
2. Eat protein for repair of damaged cells, vitamin C to fight infection and iron to prevent anaemia if blood has been lost
3. Eat easily digested foods such as white fish, chicken, custard, milk, light soups such as broth, and fruit juice
4. Take plenty of fluids especially with fever
5. Boil, bake, grill or steam food. Do not fry, reheat or add spices
6. Prepare small, attractive portions
7. Remember: hygiene is very important

 Activity 3.9 – Workbook p. 30

Exam time – Workbook p. 31

Now test yourself at *www.my-etest.com*

chapter 4

Special diets

Some people follow special diets, for many different reasons. The following cases will be studied in this chapter:
- vegetarianism
- diet-related conditions such as coronary heart disease and obesity
- eating disorders such as anorexia and bulimia
- diabetes
- coeliac disease

Vegetarianism

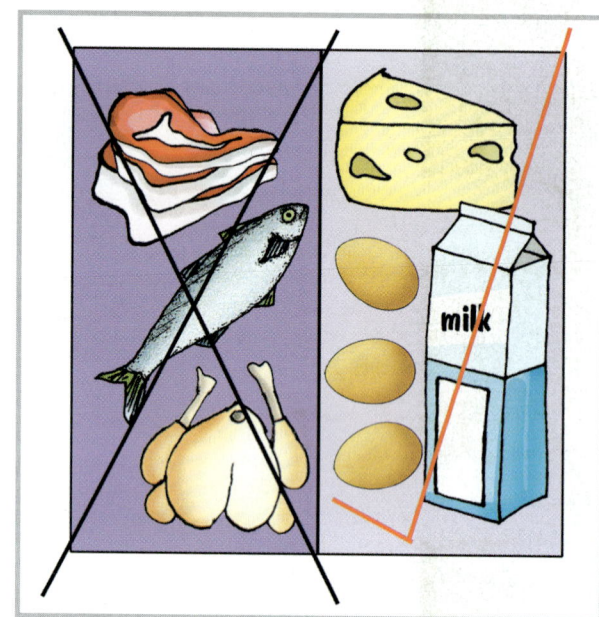

Vegetarians, for various reasons, do not eat meat or fish. There are two types of vegetarian:
- Lactovegetarians do not eat meat or fish but do eat animal products such as milk, cheese and eggs

- Vegans or strict vegetarians eat only plant foods such as fruit, vegetables, nuts, cereals and soya milk

 Activity 4.1 – Workbook p. 32

SPECIAL DIETS **4**

Case study

Hi, my name is Fiona and I'm a lactovegetarian. I became a vegetarian about three years ago. Some people do not eat meat for religious reasons; my decision was based more on health grounds. I believe it is wrong to kill animals, and also it scares me when I read about all the hormones that are pumped into animals these days. The BSE crisis is what really made up my mind for me, though.

Being vegetarian has its advantages. My diet is a really healthy one: it is low in fat and high in fibre; this I read will help prevent bowel disorders and heart disease later in life. It also keeps me at my correct weight. I do have to be careful that I get enough iron and protein though, plenty of dark green vegetables and brown bread for iron and foods like cheese, beans and TVP for protein.

Vegans usually take vitamin B supplements because this vitamin is not found in plant foods at all. Generally I find being a vegetarian fine. There are lots of different foods to choose from. I don't feel restricted just because I don't eat meat.

 Activity 4.2 – Workbook p. 32

Suitable vegetarian dishes
- various vegetable soups such as carrot, mushroom or mixed vegetable
- vegetarian pizza, quiche, risotto, lasagne, curry
- various salads
- stir-fries

Problem diets

Most diet-related problems are linked to the following:
- eating too much saturated fat (heart disease and obesity)
- eating too much salt (high blood pressure and strokes)
- eating too much sugar (obesity)
- drinking too much alcohol (heart disease, liver disease and obesity)

ESSENTIALS FOR LIVING

Coronary heart disease and strokes

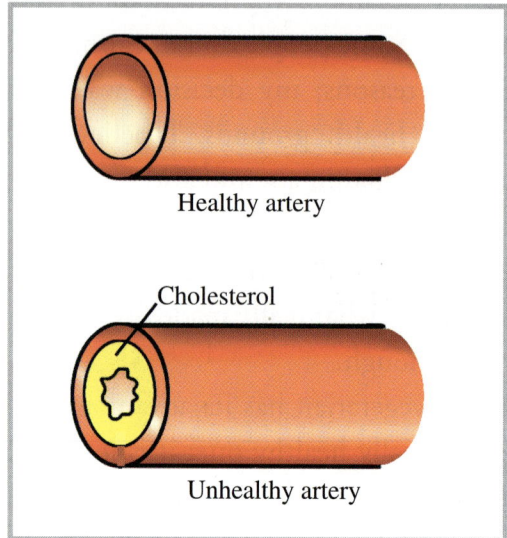
Healthy artery

Unhealthy artery (Cholesterol)

A *heart attack* occurs when an artery becomes totally blocked by the fatty substance known as cholesterol. A *stroke* occurs if the blocked artery is in the brain.

Ireland has the highest death rate from heart disease in people under sixty-five in the EU (see table below).

Most heart attacks happen within a few hours of the symptoms first occurring and before the person reaches hospital; prevention is therefore vitally important.

Heart disease is linked to:
- high-fat diets
- being overweight
- smoking
- drinking too much

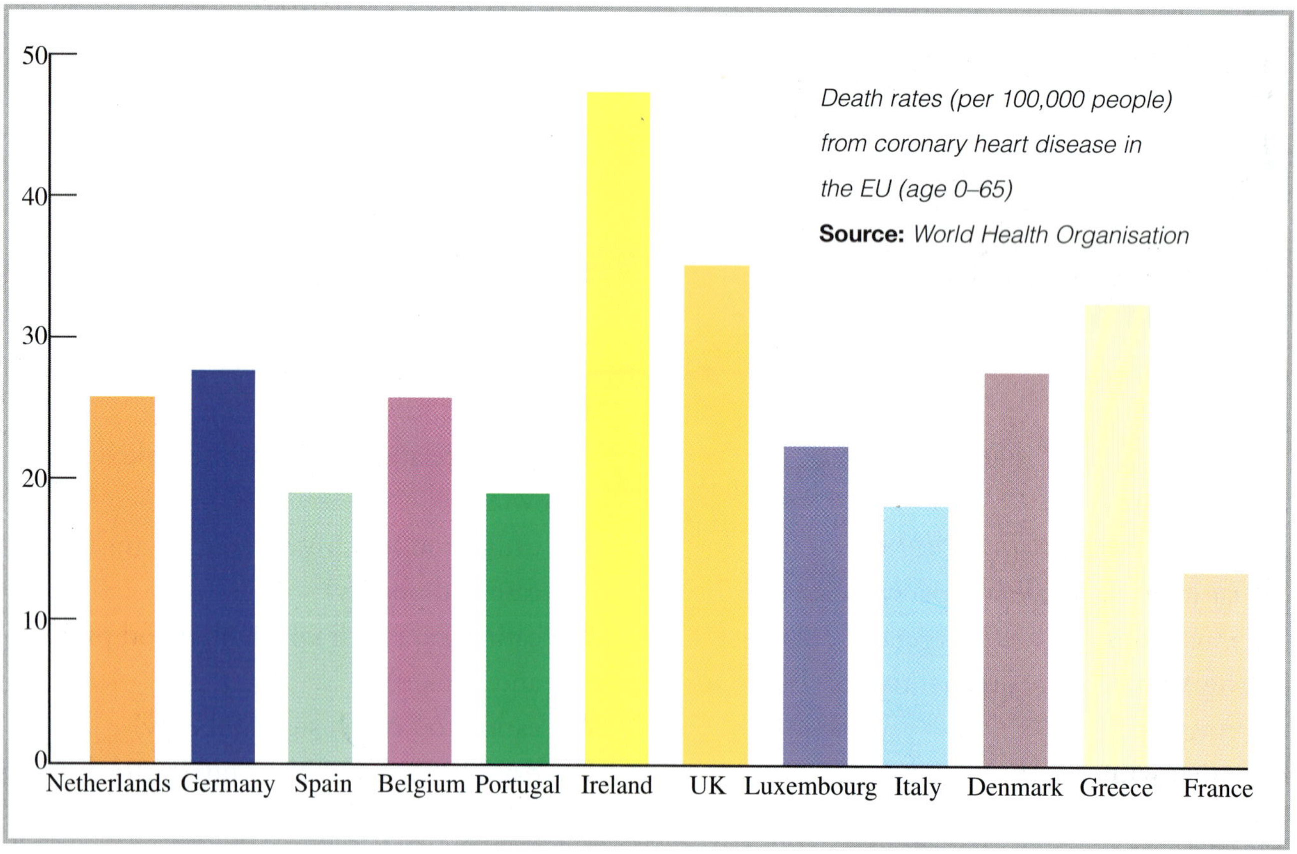

Death rates (per 100,000 people) from coronary heart disease in the EU (age 0–65)
Source: World Health Organisation

Country	Death rate
Netherlands	26
Germany	28
Spain	19
Belgium	26
Portugal	19
Ireland	47
UK	35
Luxembourg	22
Italy	18
Denmark	27
Greece	32
France	13

SPECIAL DIETS | 4

Tips for a healthy heart

Low-fat diet

Low-salt diet

Salt is the common name for a substance called sodium. Sodium is sometimes represented by the letters *Na*. Adults need about 2 g of sodium per day. Many of us, however, take up to 20 g. Too much salt in the diet can cause high blood pressure (hypertension), which can lead to heart attacks and strokes.

 Activity 4.3 – Workbook p. 33

Low-sugar diet

Sugar is not needed in the diet. Sugar and sugary food and drinks are often called *empty-calorie foods*. This means that they provide a lot of 'calories' but very little goodness.

Unfortunately, many of us are encouraged to develop a 'sweet tooth' as children; we are rewarded with sugary foods for being 'good'. Too much sugar in the diet causes weight problems and tooth decay.

 Activity 4.4 – Workbook p. 33

Obesity

ESSENTIALS FOR LIVING

Obesity means being twenty per cent or more above the normal weight for your height and build. In Ireland twenty per cent of men and sixteen per cent of women are obese.

The thyroid gland in the neck controls the rate at which we burn food. People who are naturally thin burn food quickly, but some people burn food more slowly and are therefore prone to put on weight.

Obesity can be caused by an unhealthy thyroid gland. More often, though, the cause is overeating and lack of exercise.

Problems caused by obesity:

Increased risk of:
- high blood pressure
- heart attacks
- strokes
- diabetes
- chest infections
- varicose veins
- fertility problems
- problems when under anaesthetic
- low self-esteem

Guidelines for *sensible* weight loss

1. Exercise more
2. Try not to 'snack' between meals
3. Eat less butter, red meat, sugary foods such as biscuits and chocolate
4. Eat more fruit and vegetables, white fish and chicken
5. Eat bread, potatoes and nuts in moderation
6. Drink more water

 Activity 4.5 – Workbook p. 34

COELIAC DISEASE

Higher level

During normal digestion, long protein chains are broken down into lots of tiny amino acid units. These single units pass easily through the walls of the small intestine and into the bloodstream.

Gluten is a type of protein found in wheat and some other cereals. A person suffering from coeliac disease cannot break gluten down. The big undigested gluten molecules damage the walls of the intestine as they pass through. Coeliac disease or gluten intolerance is most common in babies and young children.

The symptoms of the condition are pains in the abdomen, diarrhoea, failure to thrive and anaemia.

There is only one way of treating coeliac disease: avoid all foods containing gluten.

Foods to avoid

- all wheat products: bread, pasta, cakes, biscuits, many breakfast cereals
- processed meat products such as sausages, fish or anything else in batter or breadcrumbs
- packet soups and sauces, many snack foods

Gluten free symbol

SPECIAL DIETS 4

Gluten-free foods

- fruit, vegetables, dairy products such as milk and cheese
- rice, soya products such as TVP
- plain meat and fish
- maize products such as cornflakes
- specially made gluten-free products such as bread and cakes

Gluten-free bread

Activity 4.6 – Workbook p. 34

DIABETES

Higher level

The pancreas is a leaf-shaped organ found under the stomach. The pancreas produces a hormone called *insulin* that we need to control the amount of sugar in the blood. The pancreas of someone suffering from diabetes does not produce enough insulin.

There are two types of diabetic:

1. Insulin dependent: This is the rarest and most severe form of diabetes and can happen at any age. Twenty per cent of diabetics have this form of the illness. People with this form of diabetes must inject with insulin every day and follow a strict (low-sugar) diet. If their blood sugar goes too low they can go into a coma.
2. Non-insulin dependent: This form frequently occurs in older people who are overweight. They do not have to inject insulin every day but must follow a low-sugar diet and try to lose weight. Symptoms of this type of diabetes are thirst, excessive urine production and feeling run down.

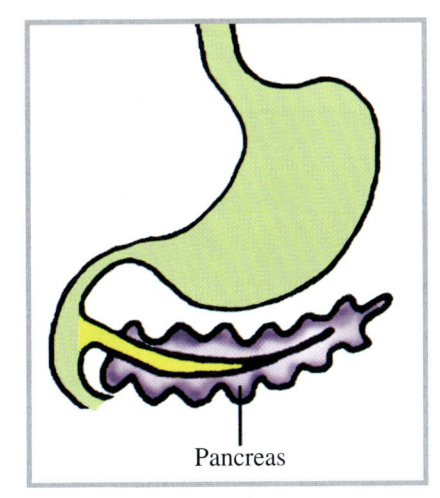

Pancreas

Dietary guidelines for diabetics

- Eat regularly: don't skip meals
- Keep to a low-sugar diet
- Eat foods high in fibre and starch

Diabetic foods such as diabetic chocolate are available.

Activity 4.7 – Workbook p. 35

ESSENTIALS FOR LIVING

EATING DISORDERS

Does the fashion world put pressure on people to be thin?

Anorexia nervosa

Anorexia nervosa is a mental illness that is most common among teenage girls. The sufferer does not have a good self-image and imagines that he or she is overweight. Anorexics starve themselves and often secretly vomit up what they do eat. This condition can be life-threatening if not treated. Treatment usually takes place in hospital and is both physical (encouraging the patient to eat) and psychological (counselling).

Bulimia

The sufferer frequently binge-eats and then vomits up or uses laxatives to get the food out of their system. Sufferers can often be of normal weight.

Activity 4.8 – Workbook p. 35

Exam time – Workbook p. 36

Revision crossword – Workbook p. 38

Now test yourself at *www.my-etest.com*

chapter 5

Meal planning

Factors that affect meal planning

1. Dietary restrictions, e.g. coeliac, diabetes, low-fat

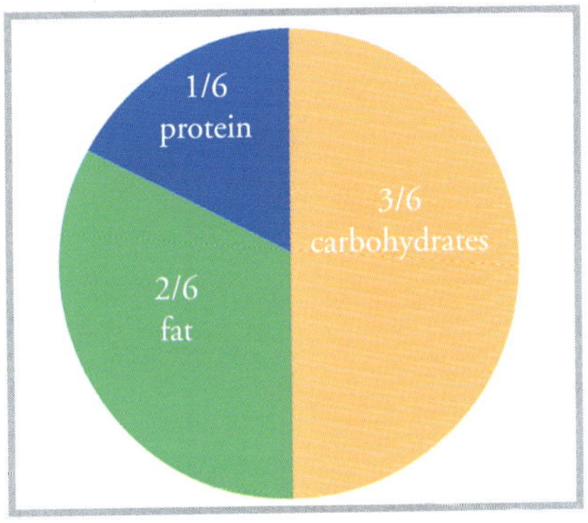

2. Meals should be nutritious and well balanced

   (1/6 protein, 2/6 fat, 3/6 carbohydrates)

3. Amount of time available to shop and cook

4. Money/equipment available

5. Climate and the availability of foods in season can also be a factor in meal planning:
   - cold weather: hot dishes such as stew
   - hot weather: cold dishes such as salads

6. Ability of the cook

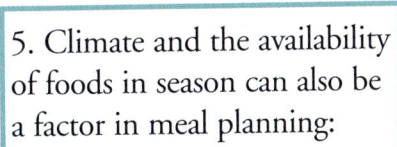

MEALS: COURSES

Main meals are usually divided into courses. In medieval times when banquets lasted for days, meals had as many as twelve courses. In modern times two to five courses are more usual.

Courses

1. Starter or appetiser
This course should be something tasty and not too filling such as soup, melon, fruit juice or pâté. This course gets the digestive juices flowing.

2. Main course
A meat, fish or vegetarian dish is central to this course. *Accompaniments* are also served, such as
- potatoes, pasta, rice
- vegetables or salad
- perhaps a sauce or gravy

3. Dessert or cheese board
- hot or cold desserts such as trifle, cheesecake, warm apple pie with ice cream
- a cheese board is a selection of three or four different cheeses served with crackers
- tea or coffee

MENUS

There are basically two types of menu:
- table-d'hôte
- à la carte

Table-d'hôte

Starters
Pan-fried garlic mushrooms
Bruchette
Toasted ciabatta bread topped with tomato, basil and olive oil
Deep-fried brie with red onion relish

Main course
Pork medallion
Medallions of pork fillet, pan-fried with julienne of onion, peppers and mushroom in a pepper sauce

Poached salmon
Fillet of salmon poached in a light cream sauce

Vegetable risotto

Chicken Piri Piri
Char-grilled corn-fed chicken fillet in lemon and chilli sauce

Desserts
Meringue nest with fresh fruit
Profiteroles with hot chocolate sauce
Hot apple pie served with fresh cream or ice cream

Tea or coffee
Price €28.00

Table-d'hôte is a set-price menu, usually cheaper than the à la carte menu, although you pay full menu price even if you do not have all the courses. Two to five courses are usually on offer, with a limited number of dishes to choose from for each course.

MEAL PLANNING

A la carte
Starters

1. Pan-fried garlic mushrooms	4.50
2. Home-made garlic bread with cheese	3.75
3. Bruchette	3.75
Toasted ciabatta bread topped with tomato, basil and olive oil	
4. Tiger prawns in filo pastry	8.25
Tossed in garlic butter and served with a sweet and sour dip	
5. Deep-fried fritter of brie with red onion relish	6.25
6. Greek salad	6.00
Fresh vine tomatoes with peppers, onions, cucumber with feta cheese and oregano dressing	

A la carte: each item on this menu is priced separately. The menu may be several pages long, with a separate page with lots to choose from for each course. Above is a sample starter page.

Rules for writing a menu
- Write down the centre of the page
- List courses in the order they are eaten
- Write the main dish of the course first and then the accompaniments
- Leave a line or place a motif between courses
- Describe dishes in some detail: give the cut of meat or fish and the cooking method, such as 'Roast loin of pork'

Activity 5.1 – Workbook p. 39

SETTING THE TABLE

- Everything should be spotless: cutlery, glasses, table mats, tablecloths etc.
- Fill clean salt and pepper containers (condiments)
- Keep flower arrangements low
- Place a jug of iced water on the table just before the meal begins

Formal table setting

Formal table setting

- Cutlery used first is on the outside
- When serving, serve food to the left of the person eating; clear empty plates from the right

Buffets

When catering for large numbers, such as a twenty-first birthday party, a buffet is often best. Food should be laid out in a logical way with drinks served at a separate table.

ESSENTIALS FOR LIVING

Hot main course	Salads	Desserts	Other
Beef, vegetable or chicken curry; pizza, quiche, cocktail sausages, chilli con carne, chicken à la king	Cold meat salads e.g. ham, turkey slices; Waldorf salad, pasta salad, potato salad, rice salad, coleslaw	Cheesecake, fresh fruit flans, mousse, fruit salad and cream, apple and rhubarb crumble or tart	Boiled rice, rolls, bread, salad dressings

A buffet table

Activity 5.2 – Workbook p. 41

PRESENTING FOOD

Before we taste food we see it; how food is presented is very important. All tableware must be *spotlessly clean*. Wipe spills from dishes or plates before bringing them to the table. Garnish or decorate food attractively, but don't overdo it.

Garnishing and decorating food

Garnish → savoury foods
Decorate → sweet foods

A garnished savoury dish

Lemon or lime loop
Lemon/lime twist
Spring onion tassel
Tomato roses
Julienne strips
Herbs

Garnishes

Garnishes

Garnishes can take the form of either (i) items placed on or beside food, such as lemon twists, or (ii) sauces, such as fillet steak with peppered sauce.

Dish	Suggested garnish
Soup	parsley, swirl of cream, croutons
Fish	lemon wedges or twists, tartar sauce
Lamb	mint sprigs, mint sauce
Turkey	cranberry sauce
Beef	horseradish
Pork	apple sauce
Duck	orange sauce, orange twists

Decorated dessert with fruit coulis

Sweet dishes are decorated in lots of ways:
- piped cream
- pastry decorations such as leaves
- fruit such as strawberry fans or slices, cherries etc.
- sugar-based decorations such as angelica, grated chocolate, hundreds and thousands
- icing, such as feathered glacé icing, royal icing on Christmas cake, dusting with icing sugar on sponge cake

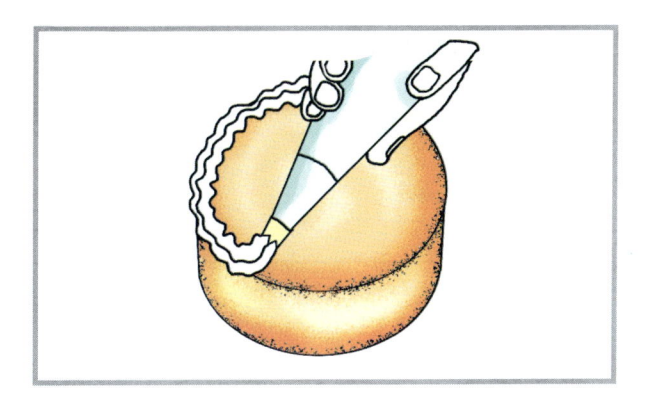

Piped cream

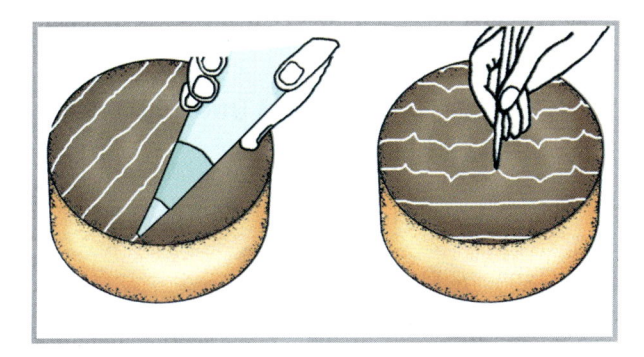

Glacé icing (feathering)

Strawberry fans

Exam time – Workbook p. 41

Now test yourself at *www.my-etest.com*

chapter 6

Good food hygiene and storage

FOOD SPOILAGE

Food spoilage is when food goes rotten or bad.
Causes: enzymes
micro-organisms (moulds, yeast and bacteria)

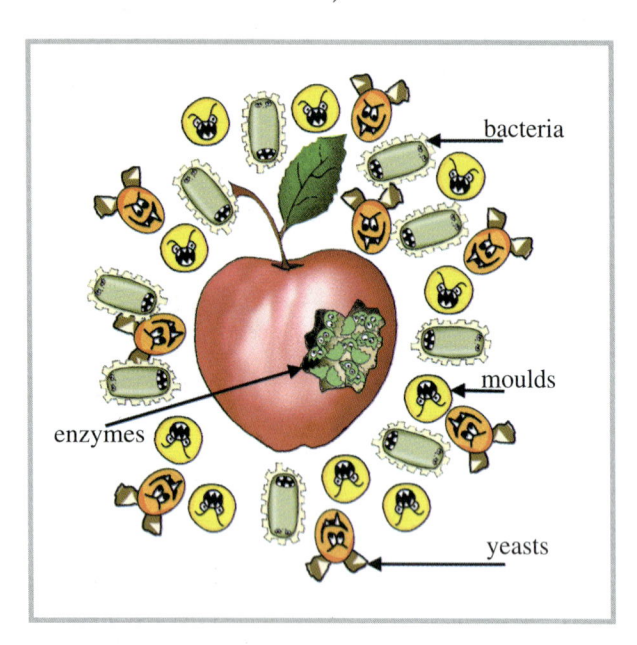

Enzymes: Enzymes are naturally present in fruit and vegetables. They cause them to ripen and eventually rot.
Moulds: Moulds cause a fluffy beard to grow on bread, fruit and vegetables. Many moulds are useful: they are used to make the antibiotic penicillin and blue-veined cheeses such as Irish blue.
Yeast: While yeast will attack some foods, such as jam, yeast has a positive use in bread-making, brewing and wine-making.
Bacteria: Bacteria are everywhere. If they multiply too much they cause food to go off. They can also cause food poisoning.

FOOD POISONING

Higher level

Food poisoning occurs when bacteria multiply to unacceptable levels on food and then the food is eaten. Our bodies respond to this invasion with symptoms such as cramps, nausea, vomiting and diarrhoea.

GOOD FOOD HYGIENE AND STORAGE **6**

Some food poisoning bacteria

Salvanella

- occurs naturally in the intestines of animals and humans; found in human and animal faeces
- causes no problem until faeces (even tiny amounts) get onto food and food is not cooked, or not cooked enough, to kill the salmonella

Staphylococci

- often found in nose, throat and skin of humans, especially on boils and sores
- when people sneeze or cough over food or handle it with uncovered sores, they risk causing this form of food poisoning

Listeria

- likes to grow on foods such as soft cheese, pâté and coleslaw
- likes cold temperatures; can multiply even in the fridge
- babies, pregnant women and the elderly are most affected; they should avoid these foods

Clostridium botulinum

- this form of food poisoning, called botulism, is very serious and thankfully very rare: the bacteria attack the nervous system, which can cause paralysis
- C. botulinum hates oxygen, so is found in faulty tins of food

Activity 6.1 – Workbook p. 43

E. coli

E. coli is the name given to a large family of bacteria commonly found in the intestines of humans and animals. Most E. coli are harmless, but one, *E. Coli 0157*, which has been in the news a lot recently, can cause serious illness in humans ranging from diarrhoea to kidney failure and death.

Human infection by E. Coli 0157 has been increasing world-wide since the early 1980s. In 1996 there was a huge outbreak in Scotland that infected over 500 people and killed twenty-one of them. The number of cases in Ireland is increasing year by year, from eight, including one death, in 1996, to seventy-six cases in 1998 (Eastern Health Board).

Main sources of infection in Ireland

Contaminated water: Careless disposal of farm and household sewage contaminates our rivers and lakes. Some rural water schemes, which draw their water from these same rivers and lakes, have a very basic system of water treatment through which E. Coli can survive. Urban water supplies are generally safer as the water receives more treatment and chlorine is added during treatment to kill E. Coli.

There is no overall authority in this country to control the standard of drinking water supplied to the public.

Eating contaminated food: Especially if it is not fully cooked. Mince, burgers, coleslaw and unpasteurised milk are high-risk foods.

Not washing hands after using the toilet: Children or some physically or mentally

ESSENTIALS FOR LIVING

handicapped people may need help with this to prevent infection. Crèche staff or people who work in hospitals need to take particular care.
Contact with infected farm animals or animal faeces.

Activity 6.2 – Workbook p. 43

Symptoms:
Stage 1: non-bloody diarrhoea
Stage 2: severe cramps followed by bloody diarrhoea (after two to three days)
Stage 3: kidney disease or failure (Haemolytic Uraemia Syndrome – US): approximately five per cent of patients die

Micro-organisms (germs) need five conditions to grow successfully

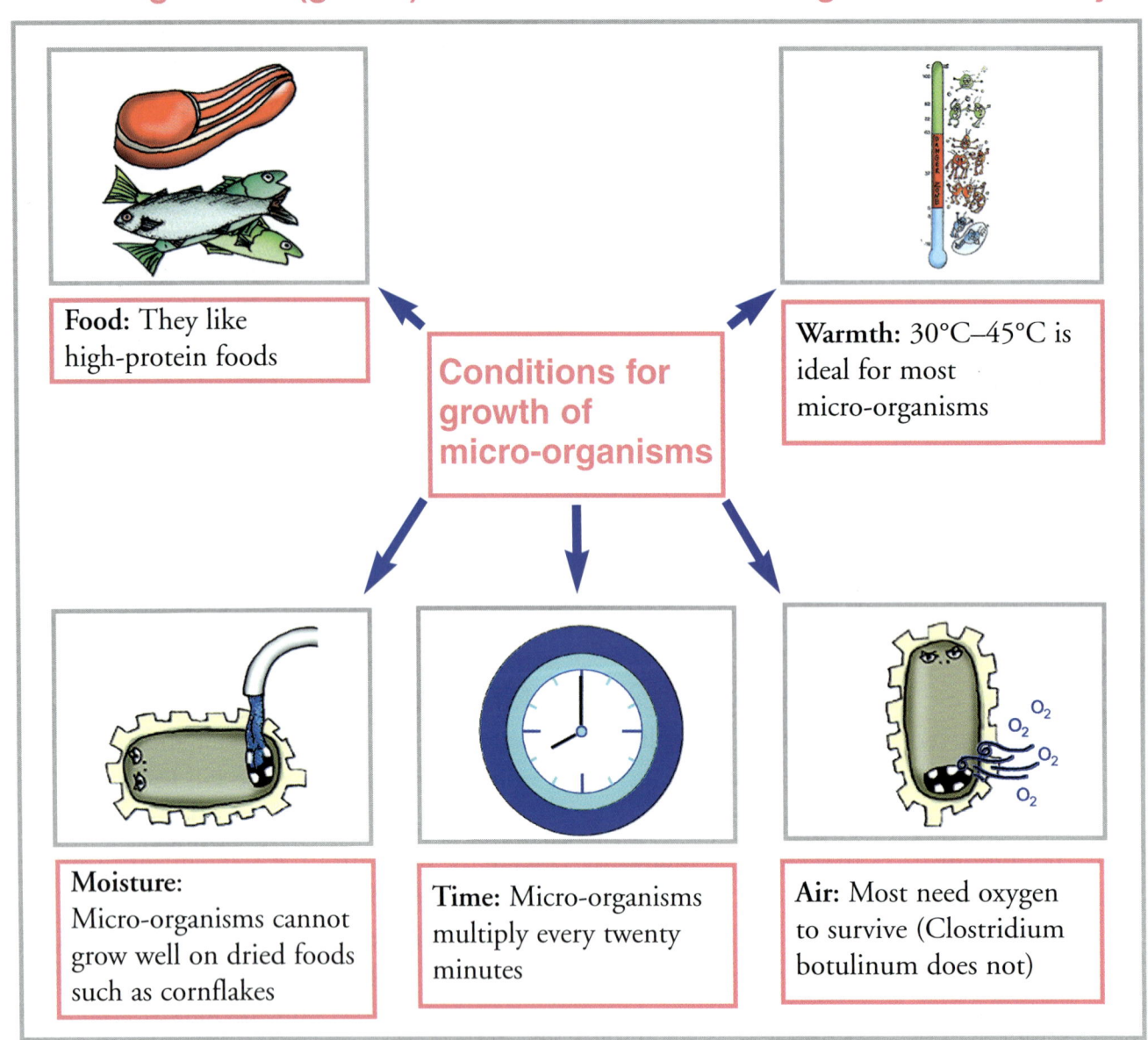

Food: They like high-protein foods

Warmth: 30°C–45°C is ideal for most micro-organisms

Conditions for growth of micro-organisms

Moisture: Micro-organisms cannot grow well on dried foods such as cornflakes

Time: Micro-organisms multiply every twenty minutes

Air: Most need oxygen to survive (Clostridium botulinum does not)

Activity 6.3 – Workbook p. 43

GOOD FOOD HYGIENE AND STORAGE   6

How food is infected by bacteria: four main ways

1. Unhygienic people: Not washing hands especially after using the toilet; not covering cuts; coughing or sneezing over food

2. Pets, vermin and insects

3. Dirt and grease: Dirt on surfaces, equipment and cloths provides food for bacteria to multiply

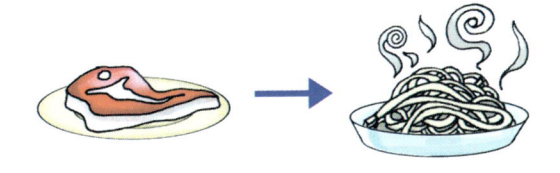

4. Cross contamination: Preparing or storing raw food, such as chicken, with food that will be eaten without being cooked or cooked again. Bacteria cross from the raw to the cooked food and this now-contaminated food is eaten, resulting in food poisoning

Activity 6.4 – Workbook p. 44

Hygiene in the kitchen

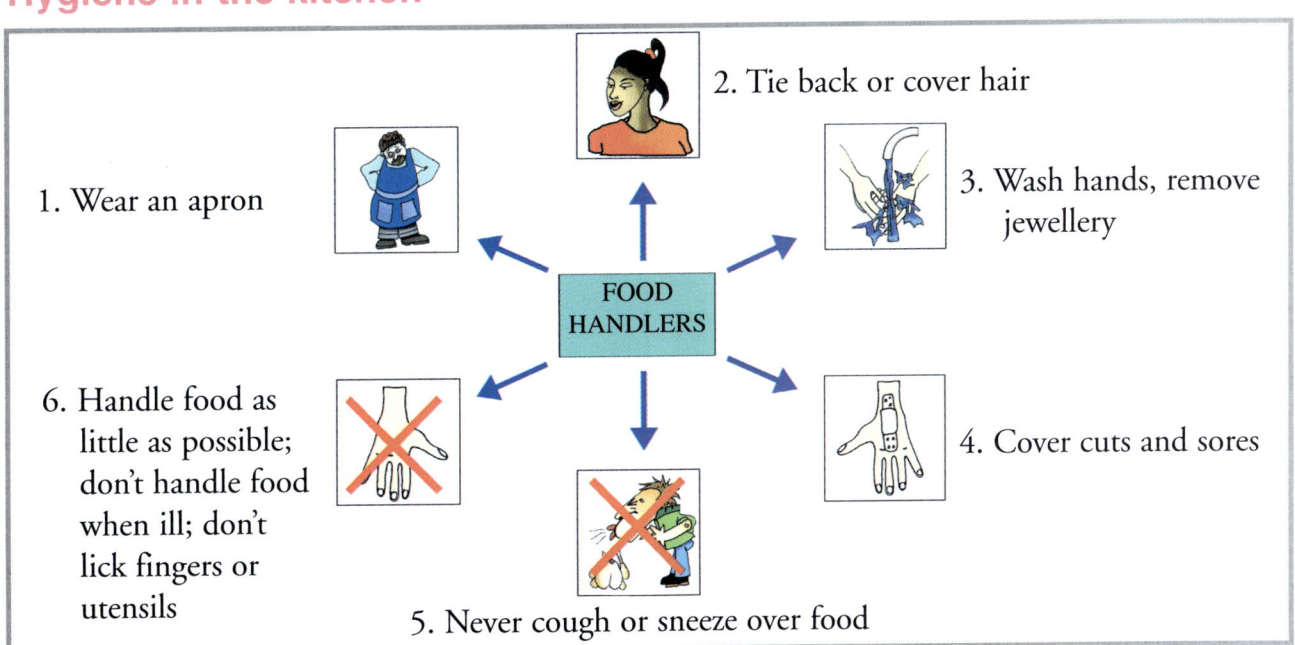

FOOD HANDLERS

1. Wear an apron
2. Tie back or cover hair
3. Wash hands, remove jewellery
4. Cover cuts and sores
5. Never cough or sneeze over food
6. Handle food as little as possible; don't handle food when ill; don't lick fingers or utensils

41

ESSENTIALS FOR LIVING

Hygiene in the kitchen continued

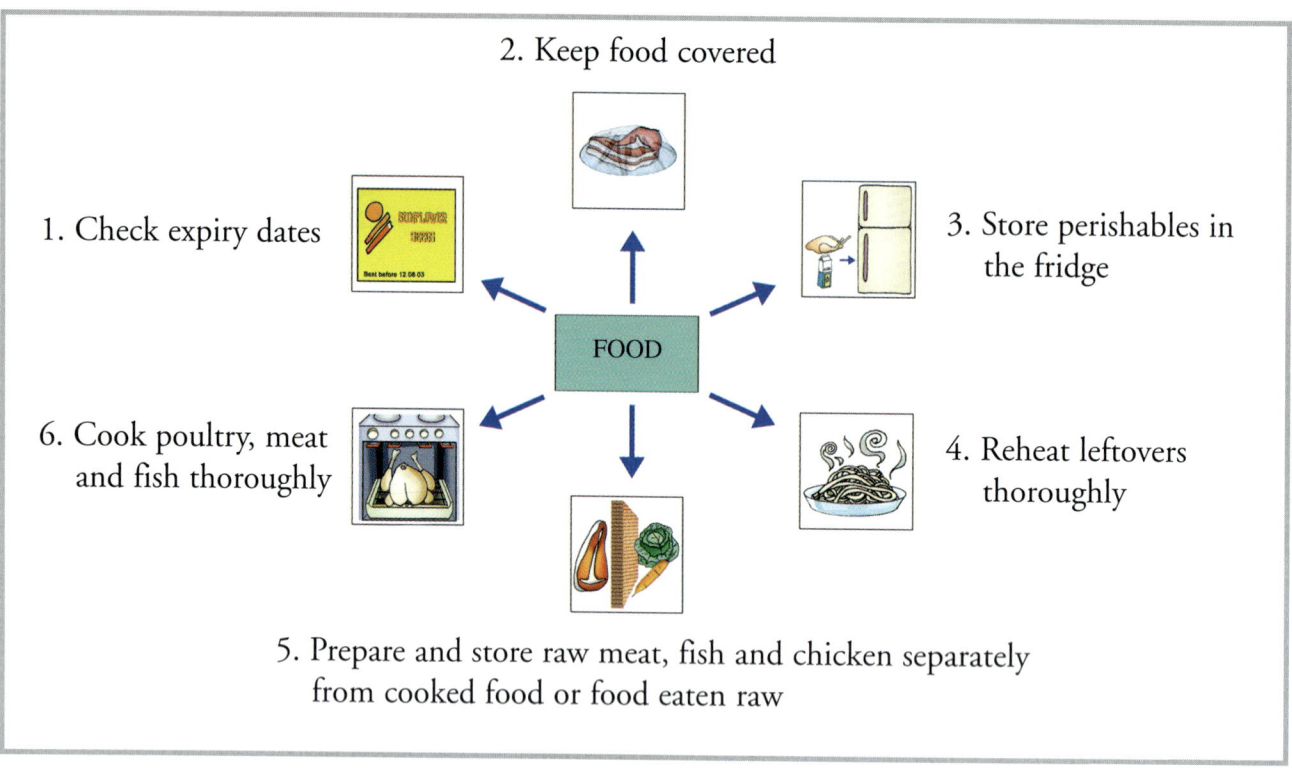

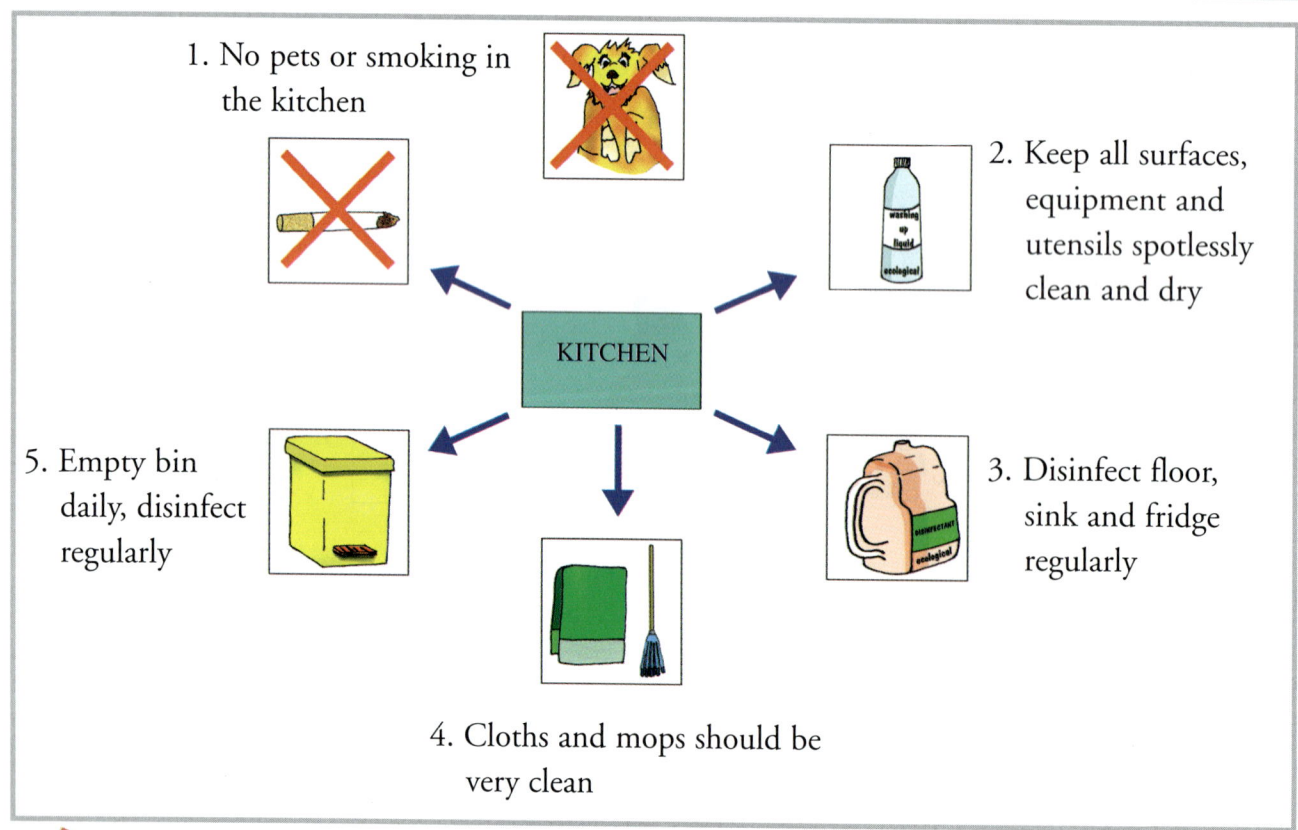

FOOD

1. Check expiry dates
2. Keep food covered
3. Store perishables in the fridge
4. Reheat leftovers thoroughly
5. Prepare and store raw meat, fish and chicken separately from cooked food or food eaten raw
6. Cook poultry, meat and fish thoroughly

KITCHEN

1. No pets or smoking in the kitchen
2. Keep all surfaces, equipment and utensils spotlessly clean and dry
3. Disinfect floor, sink and fridge regularly
4. Cloths and mops should be very clean
5. Empty bin daily, disinfect regularly

Activity 6.5 – Workbook p. 44

STORING FOOD

The *shelf life* of food (how long it remains fit to eat) depends on storing it correctly.

> Always follow the instructions on the food label.

How a food is stored depends on what type of food it is.

1. *Perishable:* Food that goes off in three or four days such as meat, milk, poultry, fish, cream and bread. Store in the fridge or freezer.
2. *Semi-perishable:* Food that lasts a week or so such as fruit, vegetables, cheese, eggs and cakes. Store fruit and vegetables in the fridge or on a vegetable rack in a cool, dark, ventilated place. Store cheese and eggs in the fridge. Store cakes etc. in an airtight tin.
3. *Non-perishable:* Food that lasts unopened for a month at least, including dried foods such as breakfast cereals, flour, rice, dried fruit, tea and tinned foods. Store in its own package in a clean, dry, wall-mounted press. Once opened store dried food in an airtight container. Treat tinned food as perishable once opened.
4. *Frozen at –18°C or below:* Many perishable and some semi-perishable foods are frozen to prolong their shelf life.

Star markings

Star markings are found on frozen food and on freezers. They tell us how long a food can be stored for:

- * one week
- ** one month
- *** three months
- **** one year

Activity 6.6 – Workbook p. 45

ESSENTIALS FOR LIVING

Packaging: for food cooking and storage

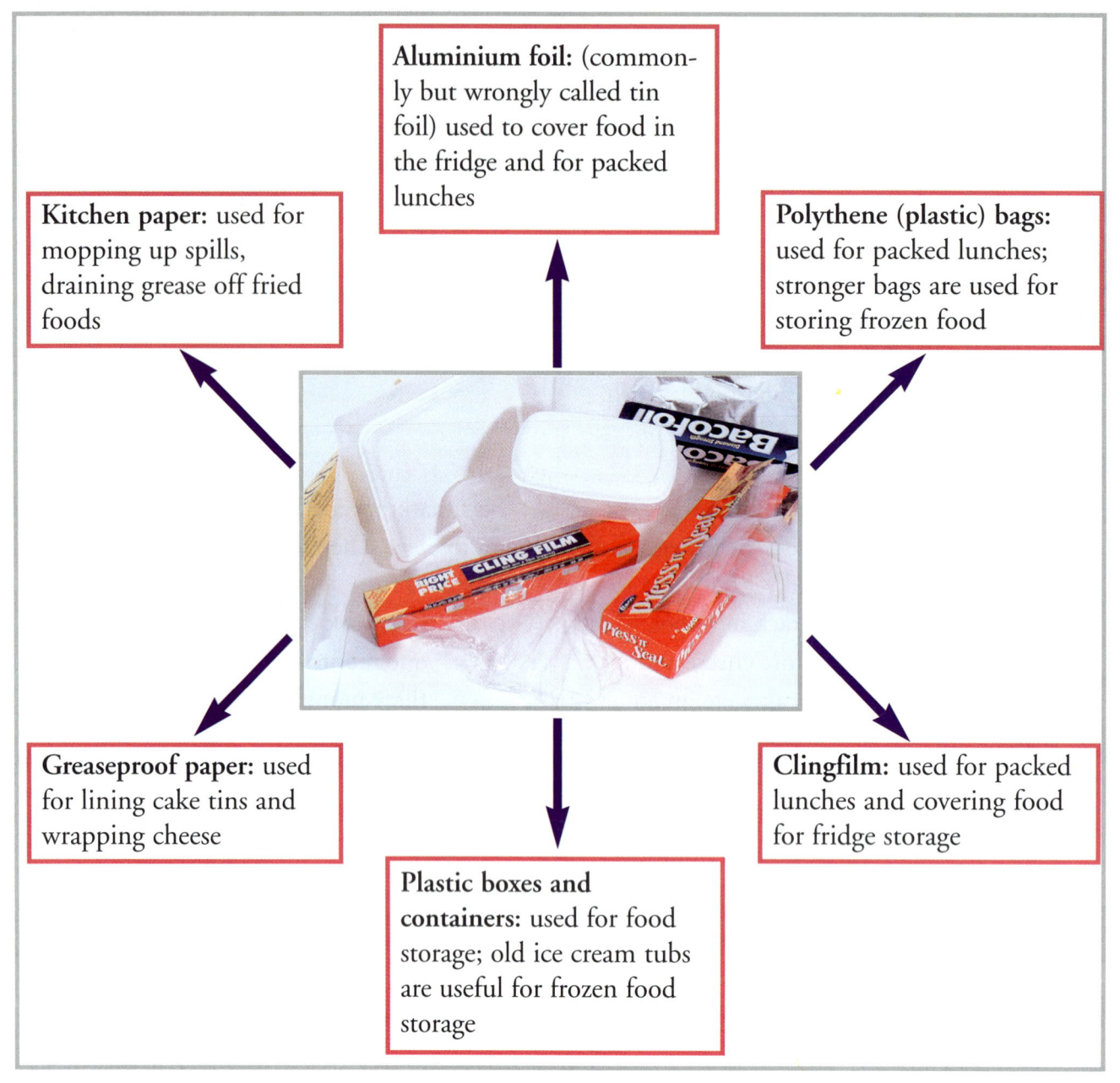

Aluminium foil: (commonly but wrongly called tin foil) used to cover food in the fridge and for packed lunches

Kitchen paper: used for mopping up spills, draining grease off fried foods

Polythene (plastic) bags: used for packed lunches; stronger bags are used for storing frozen food

Greaseproof paper: used for lining cake tins and wrapping cheese

Clingfilm: used for packed lunches and covering food for fridge storage

Plastic boxes and containers: used for food storage; old ice cream tubs are useful for frozen food storage

Activity 6.7 – Workbook p. 45

Exam time – Workbook p. 46

Revision crossword – Workbook p. 48

Now test yourself at *www.my-etest.com*

44

chapter 7

Preparing to cook

KITCHEN SAFETY

Most household accidents occur in the kitchen. When in the kitchen: take care, be safe.

Kitchen safety guidelines

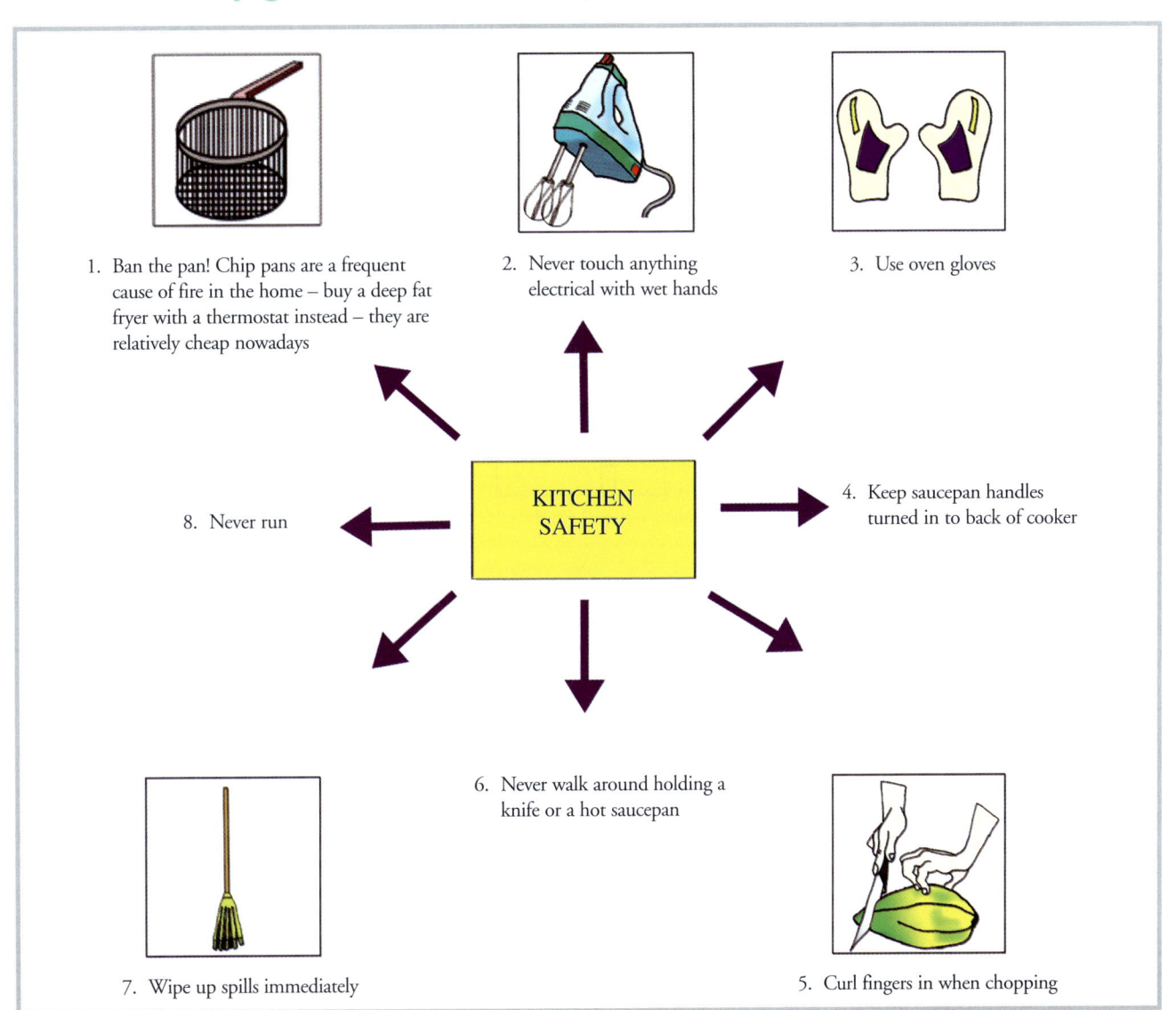

1. Ban the pan! Chip pans are a frequent cause of fire in the home – buy a deep fat fryer with a thermostat instead – they are relatively cheap nowadays
2. Never touch anything electrical with wet hands
3. Use oven gloves
4. Keep saucepan handles turned in to back of cooker
5. Curl fingers in when chopping
6. Never walk around holding a knife or a hot saucepan
7. Wipe up spills immediately
8. Never run

45

Before cooking

Follow this work routine before you start to cook:

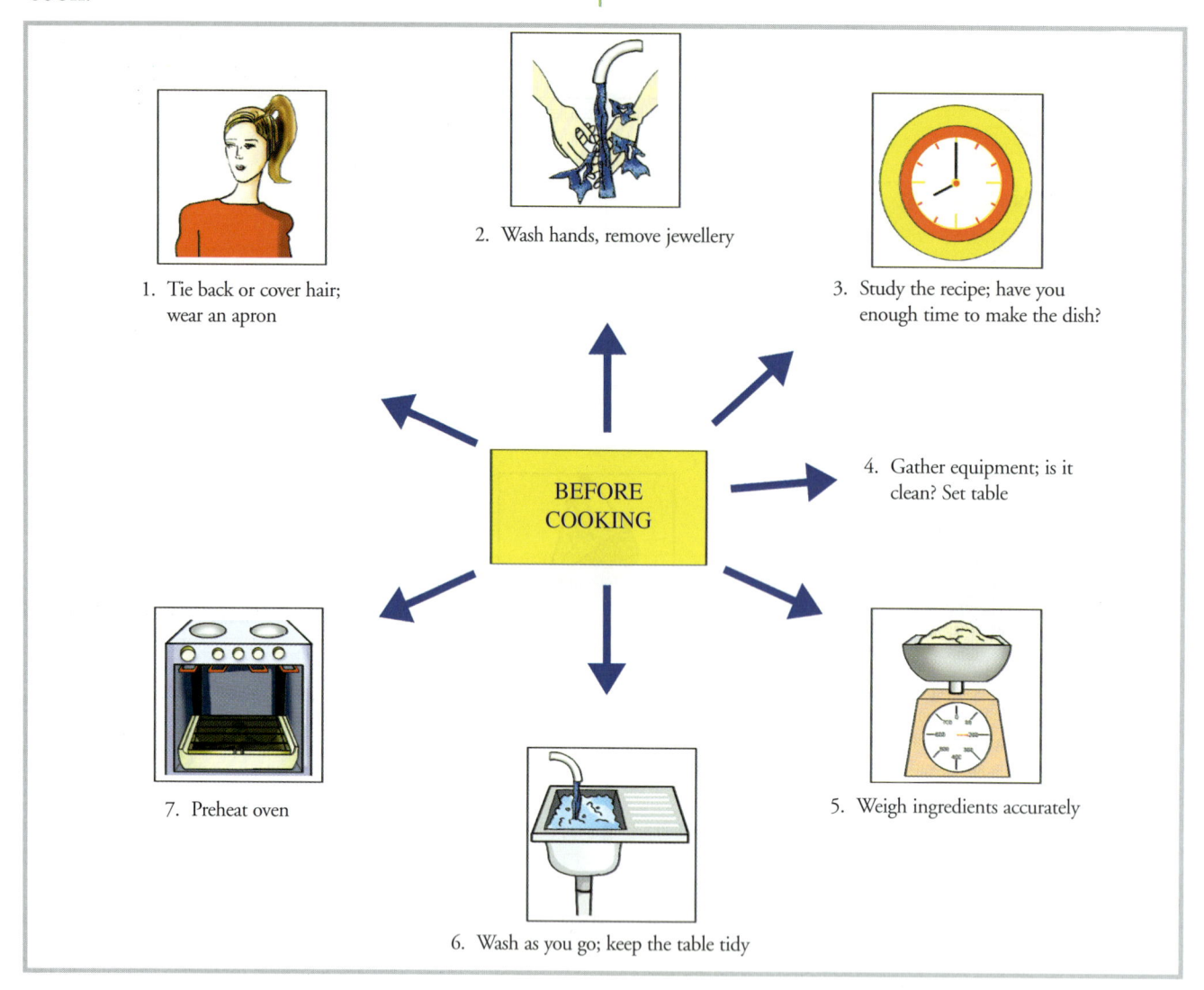

1. Tie back or cover hair; wear an apron
2. Wash hands, remove jewellery
3. Study the recipe; have you enough time to make the dish?
4. Gather equipment; is it clean? Set table
5. Weigh ingredients accurately
6. Wash as you go; keep the table tidy
7. Preheat oven

READY TO GO

A recipe has:
- the name of the dish
- sometimes a picture of the finished dish
- a list of ingredients
- the method: instructions on how to make the dish
- serving suggestions
- oven temperature and cooking time

Weighing (solids) and measuring (liquids)

Solids

Solids are weighed in grams (g). As a gram is a very small amount you will rarely see less than 25 g written in a recipe. If less than 25 g is needed the recipe will usually say a teaspoon (5 g), dessertspoon (10 g) or a tablespoon (15 g). Recipes mean a level spoon unless otherwise stated. **Note:** 1000 g = 1 kg

PREPARING TO COOK   7

Weighing

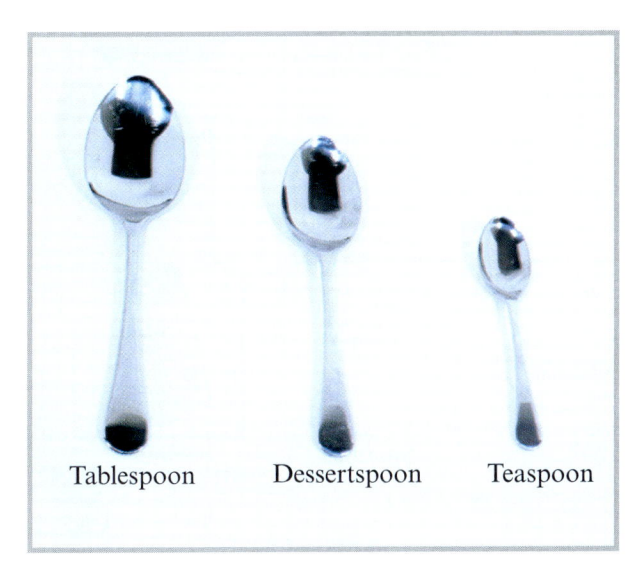

Spring scale

Balance scale

Tablespoon   Dessertspoon   Teaspoon

Spring scales: Make sure the dial is at 0 after you place the empty bowl on the scales.

Balance scales: Select the weight you need, such as 100 g. Add ingredient to the other side until scale is level or balanced.

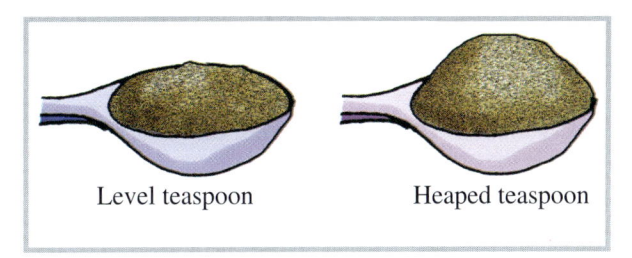

Level teaspoon   Heaped teaspoon

Recipes mean a level teaspoon unless otherwise stated

Measuring

Liquids are measured in millilitres (ml). There are 1000 ml in one litre (L). A measuring jug is used to measure liquid; again less than 25 ml is rare in recipes. Teaspoons (5 ml), dessertspoons (10 ml) and tablespoons (15 ml) can be used to measure liquids also.

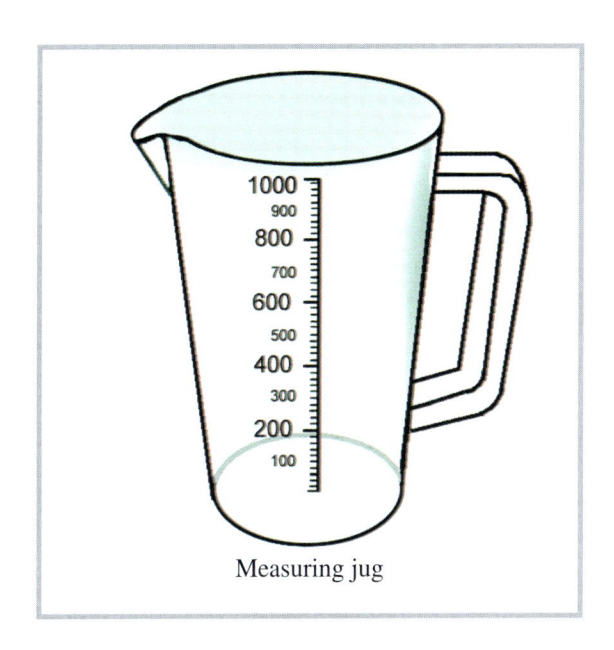

Measuring jug

47

ESSENTIALS FOR LIVING

Measuring spoons

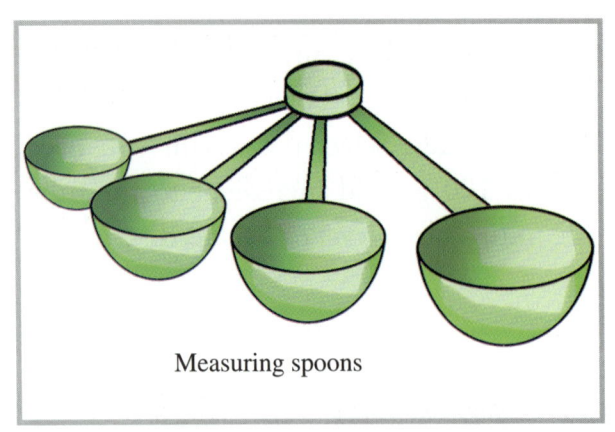

Special measuring spoons, while not essential, can be handy in the kitchen

Activity 7.1 – Workbook p. 49

Some common utensils
- bowls
- cutlery
- pie dishes
- saucepans
- baking tins
- frying pans
- casserole dishes
- chopping board
- rolling pin
- wooden spoons
- sharp knives
- potato peeler
- measuring jug

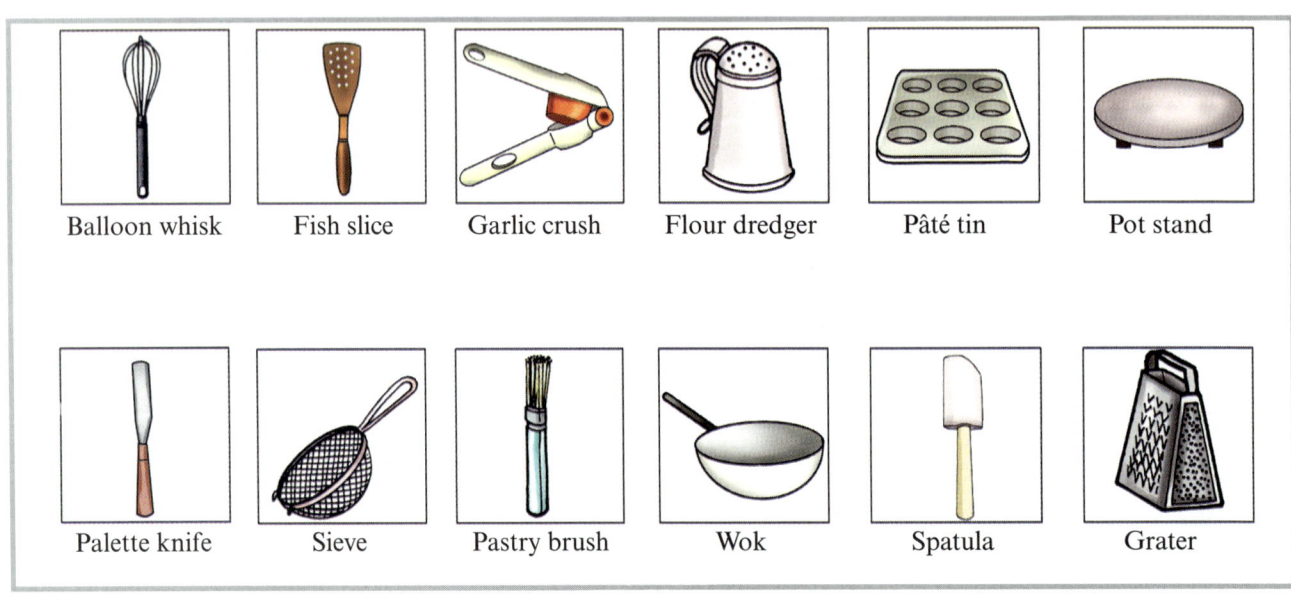

EQUIPMENT

Equipment consists of:
1. Utensils, e.g. cutlery, bowls etc.
2. Large appliances, e.g. cooker, fridge etc
3. Small appliances, e.g. mixer, food processor etc.

Utensils

Utensils are small pieces of equipment that need to be clean and well cared for.

| Balloon whisk | Fish slice | Garlic crush | Flour dredger | Pâté tin | Pot stand |
| Palette knife | Sieve | Pastry brush | Wok | Spatula | Grater |

Some less common utensils

48

PREPARING TO COOK **7**

How you clean and care for kitchen utensils depends on what they are made from: glass, plastic, wood or metal.

Cleaning kitchen utensils

Glass:	• steep to loosen food • wash in hot soapy water, rinse and dry • do not place in the oven unless 'heat resistant' glass e.g. pyrex • hot glass will crack if placed on a very cold surface: use a pot stand	Plastic:	• wash in hot, soapy water, rinse and dry well • will melt if put near direct heat
Wood:	• scrub with the grain in warm soapy water • do not store away until completely dry or wood will warp	Metal:	• wash in warm soapy water, dry thoroughly (place in a warm oven or on a radiator to dry completely) • never use abrasives e.g. brillo or metal utensils on a non-stick surface or it will be destroyed • store saucepans with the lids off

Activity 7.2 – Workbook p. 50

Large appliances
- cooker
- fridge
- microwave
- dishwasher
- freezer

Using the cooker
Fuel: cookers can run on either electricity, gas or solid fuel.

A cooker can consist of:
- a hob
- a grill (sometimes also a small top oven)
- an oven

Preheating: Ovens can take ten minutes to heat up (gas ovens heat up more quickly). With electric ovens a light usually goes off when the oven has reached the temperature it has been set at. You must preheat the oven for bread and cakes in particular.

Setting the oven: Oven temperature is written in degrees Celsius (°C) or gas mark 1 to 9. Some older ovens may be in degrees Fahrenheit (°F). Each recipe will give an oven temperature and a cooking time.

Types of oven: There are basically two different types of oven:
- conventional
- fan (fan in oven at the back)

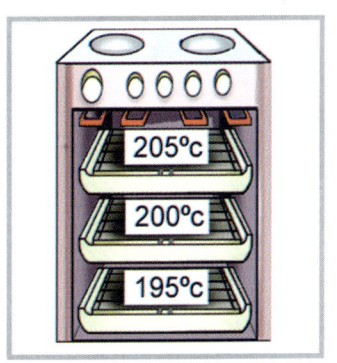

A conventional oven set at 200°C

49

ESSENTIALS FOR LIVING

The temperature in a fan oven is the same on every shelf. In a conventional oven, the top shelf is hottest, the middle shelf is as you set it and the bottom shelf is coolest.

Note: arrange the shelves before you turn the oven on. Never place food on the floor of the oven as it will burn.

Activity 7.3 – Workbook p. 50

Small appliances

- liquidiser
- hand-held electric mixer
- food processor
- electric knife

Liquidiser

Uses: blend soups, make breadcrumbs, purée cooked fruit and vegetables, no heavy duty work

Food mixer

Uses: cream sugar and margarine, whip cream, whisk egg whites, no heavy duty work

Food processor

Uses: can do everything liquidiser and mixer can do plus: shred, dice, chip raw vegetables, mix stiff doughs and cakes

Care and cleaning of small motorised appliances

Remove loose parts and wash in hot soapy water. Dry thoroughly and store without putting appliance back together. Never put motor in water, just wipe with a damp cloth. Water and electricity are dangerous together.

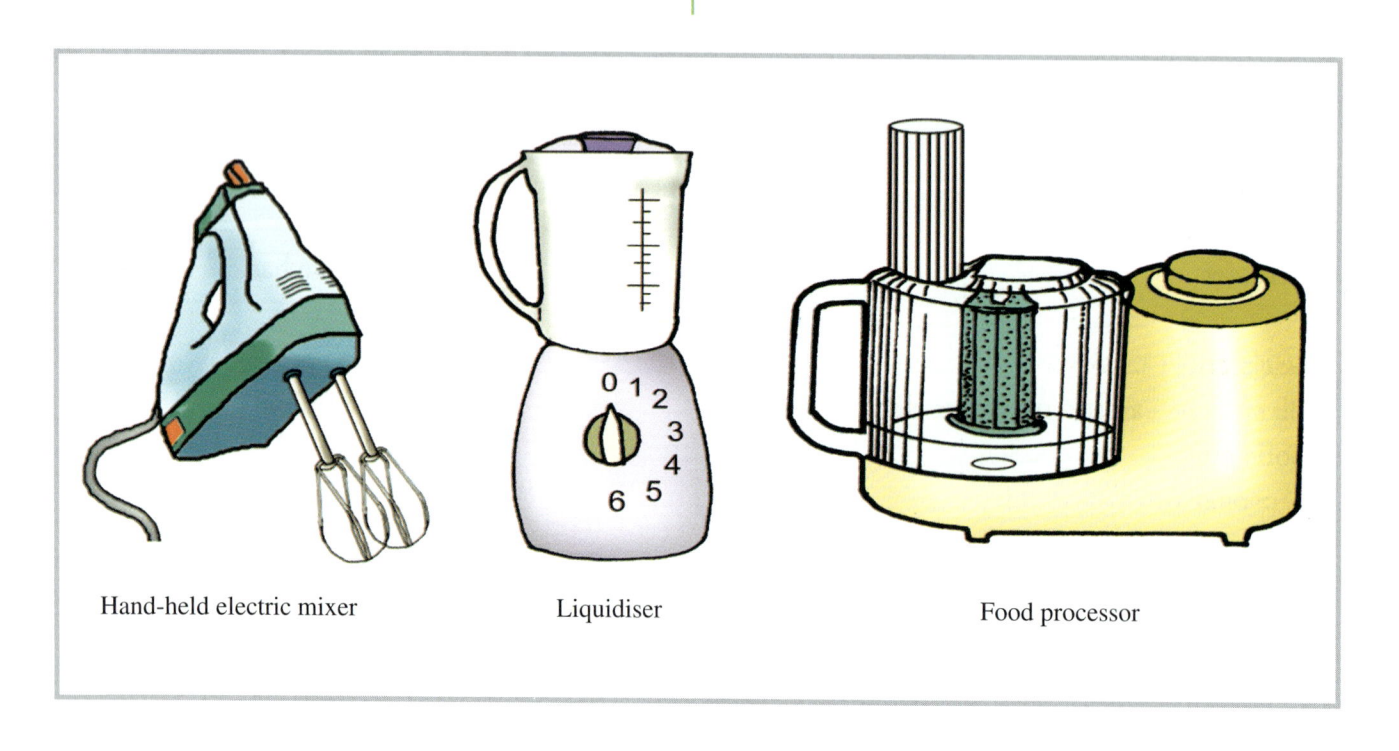

Activity 7.4 – Workbook p. 51

Hand-held electric mixer   Liquidiser   Food processor

GLOSSARY OF COOKERY TERMS

You will come across these frequently in recipes. Read through and learn the list below before completing the crossword on page 51 of your workbook.

Aerate: introduce air to a mixture (sieving, rubbing in, or whisking)

Al dente: cook food (pasta, vegetables) so they still have bite and are not really soft

Au gratin: food cooked in sauce, sprinkled with cheese or breadcrumbs, then browned under the grill or in the oven

Baste: spoon hot fat and meat juices over roasting meat to stop it drying out

Bind: bring a mixture together, for example adding beaten egg to minced meat to make burgers

Blanch: plunge foods into boiling water, then into cold: this removes skins (tomatoes and almonds) or destroys enzymes (vegetables) before freezing

Blend: gently add an ingredient to a mixture

Bouquet garni: a bunch of herbs (or a commercial sachet) added to flavour soups and stews and later removed

Brine: salty water

Coat: cover with, for example, batter, breadcrumbs or sauce

Consistency: thickness of a mixture

Cream: beat foods together until they are soft and creamy

Croutons: small cubes of fried bread used as a garnish for soups

Dredge: to sprinkle sugar or flour (for instance dredging a table with flour so that pastry doesn't stick to it while rolling)

Fold: gently adding an ingredient to a mixture

Garnish: decorating a savoury dish

Glaze: brush over bread and cakes before baking to give them a shine when cooked

Knead: to work dough with hands, for instance when making bread

Marinade: steep foods in flavoured liquid before cooking

Parboil: partially cook by boiling (potatoes can be parboiled before oven roasting)

Purée: mincing up fruit or vegetables

Raising agent: produces gas in a mixture (such as bread dough) and causes it to rise

Roux: a mixture of equal quantities of fat and flour used as a base for many sauces

Sauté: fry gently for a short time in hot fat

Season: add salt, pepper, herbs or spices to food

Shortening: fat added to bread and cakes

Simmer: cook gently just below boiling point

Texture: the feel of something: smooth, lumpy, coarse, crisp, soft etc.

Activity 7.5 – Workbook p. 51

Exam time – Workbook p. 52

Now test yourself at *www.my-etest.com*

chapter 8

Cooking food

WHY FOOD IS COOKED

- Cooking kills harmful bacteria, making food safe to eat. It also has the effect of preserving food, for example cooked chicken will last longer than raw chicken in the fridge.
- Cooking makes some foods easier to chew and digest, for example meat and starchy foods such as potatoes.
- Cooking improves the colour and flavour of many foods.

Effects of cooking on food

Food loses water and shrinks e.g. meat

Protein coagulates (sets), e.g. eggs, cheese, skin on milk

Flavours and aromas develop

Fat melts

EFFECTS OF COOKING ON FOOD

Nutrients are lost especially Vitamins C and B

Food goes soft and breaks up, e.g. potatoes, apples

Food absorbs water and swells, e.g. rice

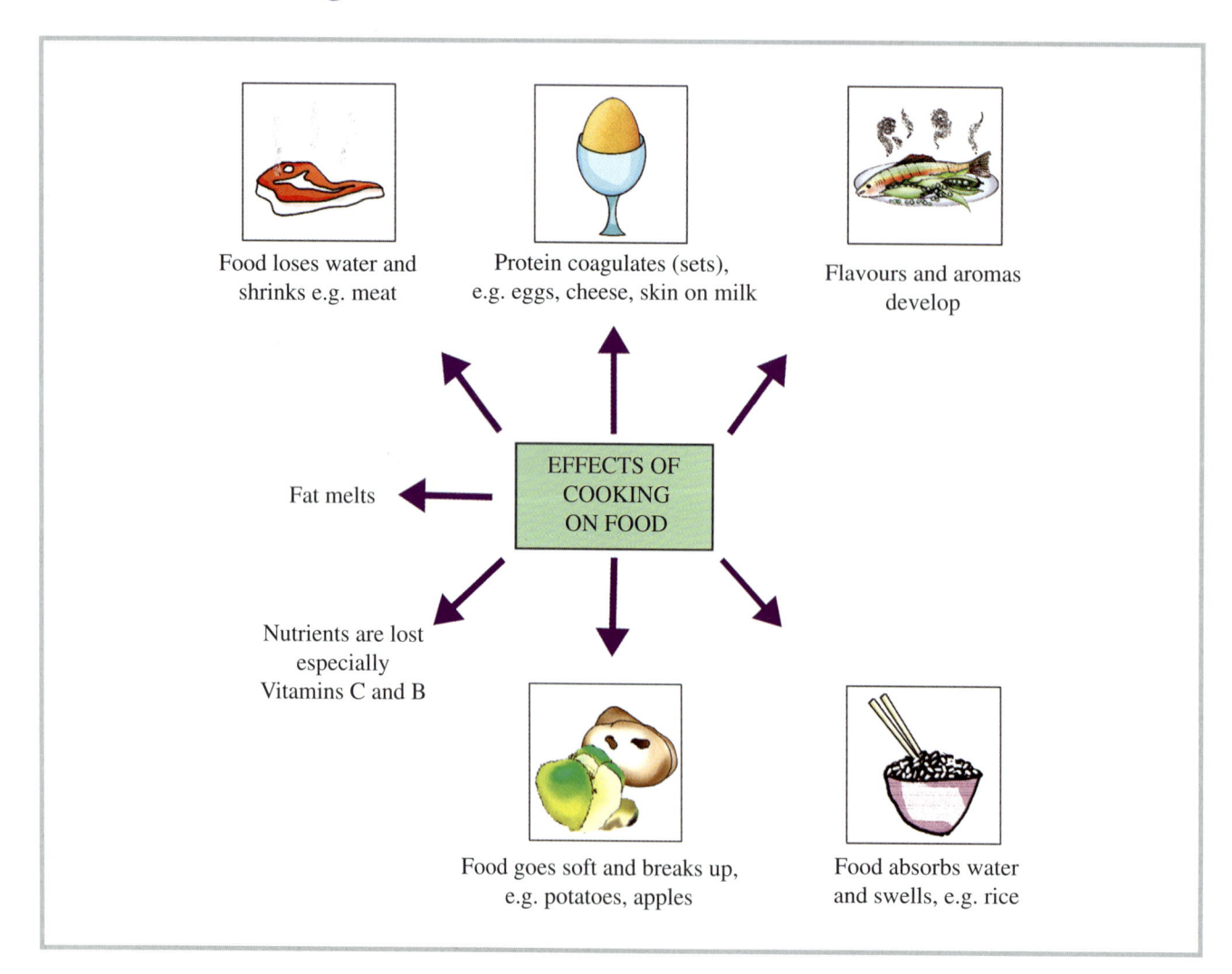

52

HOW FOOD IS COOKED: METHODS OF HEAT TRANSFER

Higher level

Conduction

Example: frying

Heat travels from molecule to molecule along a solid object. For example in frying, hot molecules in the cooker-ring touch the molecules in the frying pan and cause them to get hot; they in turn touch the food and cause it to get hot, thus cooking it.

Heat travelling by conduction

Convection

Examples: boiling, deep-fat frying, baking

When liquid and gases are heated they rise. They are replaced by cooler gas or liquids. These movements are called convection currents. These currents heat the food by convection.

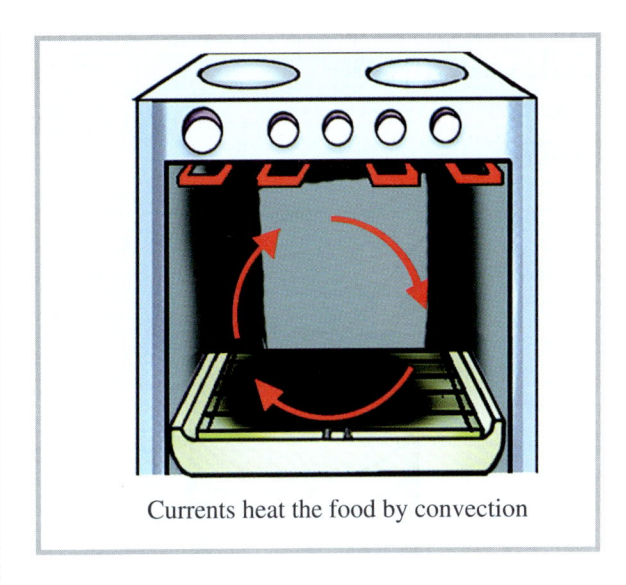

Currents heat the food by convection

Radiation

Example: grilling

Hot rays from a heat source hit food and heat its surface. The heat then travels into the middle of the food by conduction, thus cooking it through.

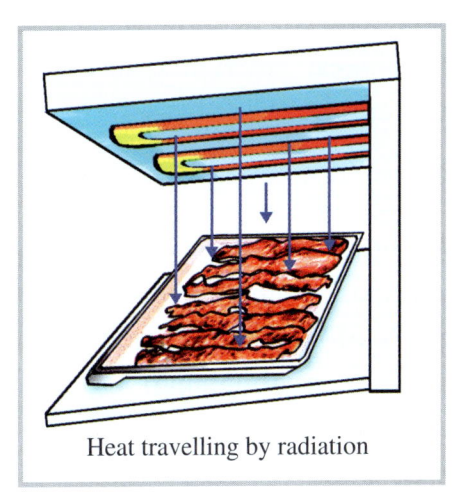

Heat travelling by radiation

METHODS OF COOKING

1. Moist methods: boiling and simmering, stewing, casseroling, poaching, steaming, pressure-cooking
2. Dry methods: grilling, baking
3. Methods using fat: frying (shallow-, deep- and stir-frying), roasting
4. Other: microwave cooking

Moist methods

Boiling

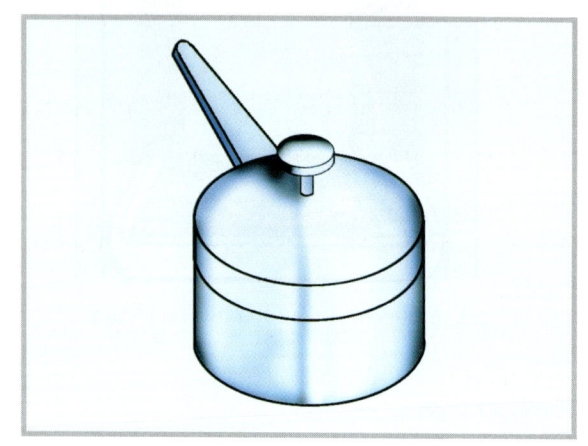

Use a heavy saucepan with a tight-fitting lid

Description: food is cooked in bubbling liquid (100°C).
Suitable foods: meat, vegetables, rice, pasta.
Advantages: economical, needs little attention, safe for beginners.
Disadvantages: nutrients lost into the cooking liquid.

Tips
- use a heavy saucepan with a tight lid
- use cooking liquid for soups and sauces
- once food is bubbling rapidly reduce to simmer (barely bubbling)

Stewing and casseroling

Description: these are slow moist cooking methods. Food is cooked in liquid in a heavy lidded saucepan on the hob (stewing), or in a dish in the oven (casseroling). The food and the liquid it has been cooked in are often both eaten.
Suitable foods: tough cuts of meat, fruit.
Advantages: economical: a whole meal can be cooked on one ring; tough, cheaper cuts of meat may be used; cooking liquid is consumed so nutrients are not lost.
Disadvantages: slow.

Tips
- simmer (don't boil) stew
- keep lid on
- use stock as a cooking liquid instead of water: it adds flavour

Poaching

Description: food is cooked gently in water that is barely simmering. A shallow saucepan or poacher is used.
Suitable foods: eggs, fish.
Advantages: low calorie method; food remains digestible (good method for invalid/convalescent cookery).
Disadvantages: foods can break up; constant attention required.

Tip: do not let water boil rapidly or food will break up and be spoiled.

Poacher

COOKING FOOD   8

Steaming

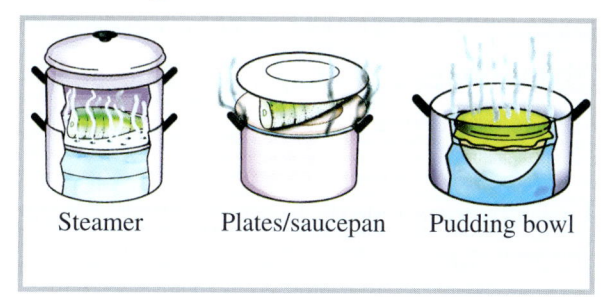

Steamer · Plates/saucepan · Pudding bowl

Description: food is cooked in rising steam.
Suitable foods: vegetables, fish, puddings.
Advantages/disadvantages: steamed food can lack flavour, but is usually very digestible.

Pressure cooking

Description: food is cooked quickly with steam under pressure.
Suitable foods: vegetables, tough cuts of meat, puddings.
Advantages: quick: pressurised steam is very hot and cooks food more quickly than boiling or steaming.
Disadvantages: can be dangerous if pressure cooker is not used correctly.

Tip: time cooking carefully; overcooking is common.

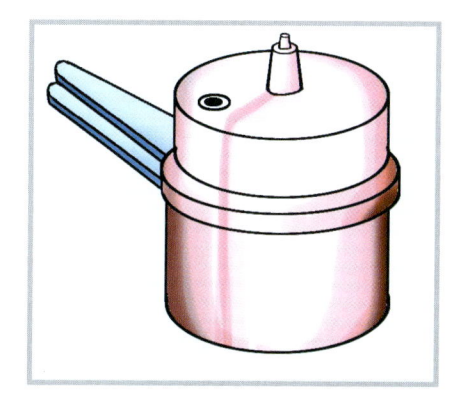

Pressure cooker

Dry methods

Grilling

Description: food is cooked by rays from a hot gas or electric grill.
Suitable foods: tender meat, e.g. steak; fish; some fruit and vegetables (tomatoes, grapefruit).
Advantages: quick, not greasy, food has good texture.
Disadvantages: only expensive meat can be used; needs constant attention.

Tip: clean grill pan after each use or old oil will smoke next time.

Baking

Description: food is cooked by dry heat in an oven.
Suitable foods: bread, cakes, tarts, pies, apples, potatoes, puddings, fish.
Advantages: needs little attention, safe for beginners, adds flavour.
Disadvantages: expensive if oven is used for one item.

Tips:
- batch-bake to make full use of the oven
- always preheat the oven fully
- adjust the shelves before the oven heats up
- don't open the oven door especially at the start of cooking time

Methods using fat

Frying

Description: food is cooked in hot fat by deep-, shallow- or stir-frying.
Suitable foods: tender meat and chicken, fish, eggs, chips, onion rings, mushrooms.
Advantages: quick, good flavour.
Disadvantages: method adds kilocalories, dangerous, greasy, needs constant attention.

Tips:

- heat fat fully before adding the food or it will soak up fat and become soggy
- fire is a risk with frying: do not leave a frying pan or wok unattended
- do not use a chip pan: use a deep-fat fryer with a thermostat instead
- drain fried food in kitchen paper to remove some greasiness

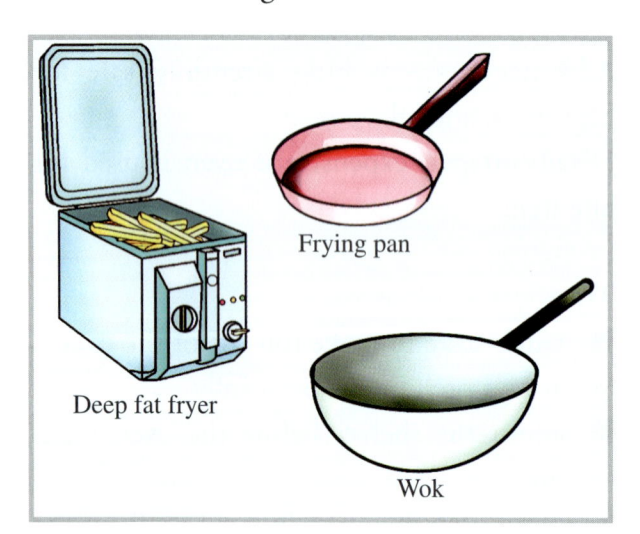

Frying pan

Deep fat fryer

Wok

Roasting

Description: food is cooked in the oven and basted with hot fat (hot fat is spooned over).
Suitable foods: meat; poultry: chicken, turkey, duck; potatoes and some vegetables e.g. parsnips.
Advantages: nice flavours, needs little attention.
Disadvantages: basting adds kilocalories; food is dry if not basted.

Tip: covering meat helps prevent drying: uncover thirty minutes before end of cooking to give brown colour.

Microwave cooking

Description: invisible 'microwaves' bounce around the inside of the oven. The 'microwaves' hit the food and cause the molecules in it to vibrate. This vibrating produces heat, which cooks the food.
Suitable foods: reheats, thaws and cooks most foods.
Advantages: quick.
Disadvantages: does not brown food.

Tips:

- do not put metal dishes or dishes with a metal rim in the microwave or sparks will fly!
- foods with a skin, e.g. tomatoes, will explode if you do not prick them with a fork first
- allow food to 'stand' after cooking before eating

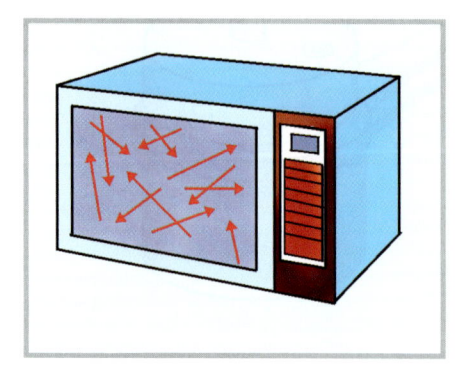

Microwave cooker

Activity 8.1 – Workbook p. 55

RECIPE MODIFICATION

Modifying a recipe means changing it in some way.

Possible reasons:
- to make the dish healthier
- to make it suitable for people with special dietary needs, e.g. vegetarians, coeliacs
- to increase or decrease the size of the dish
- to substitute available ingredient(s) for an unavailable one

Ways to modify a recipe for health reasons

↓ *Reduce fat*
- use low-fat milk, cheese etc. in recipes
- use polyunsaturates, e.g. Flora, instead of hard fats, e.g. margarine
- fry meat in its own fat on a non-stick pan
- reduce the amount of meat in a recipe and increase vegetable content
- don't use cream in desserts; use low-fat yogurt instead

↓ *Reduce sugar*
- use artificial sweeteners where possible
- avoid icing cakes, buns etc.
- serve fruit salad in orange juice instead of sugar syrup

↓ *Reduce salt*
- flavour with herbs instead
- do not add salt to baked products, e.g. scones; use salt substitutes if necessary

↑ *Increase fibre*
- leave skins on fruit and vegetables where possible
- use brown rice, pasta etc.

Activity 8.2 – Workbook p. 55

Activity 8.3 – Workbook p. 56

Exam time – Workbook p. 56

Now test yourself at *www.my-etest.com*

chapter 9

The practical cookery exam

The practical cookery exam is worth thirty-five per cent (higher level) and forty-five per cent (ordinary level) of your final Junior Certificate Home Economics mark. It is therefore very important that you prepare well and carry out the task given to you to the best of your ability.

EXAM PROCEDURE

Approximately two weeks before the exam (exams are usually held in April), you will be asked by your Home Economics teacher to draw a task from a selection of tasks.

Sample task
Nutritional surveys show that teenagers' diets are sometimes deficient in iron. Bearing this in mind, design and set out a menu suitable for a main meal for teenagers. Prepare, cook and serve the main course dish. Prepare and serve a suitable accompanying dish or salad.

Food and Culinary Skills Tasks 2002

Once you have drawn your task you can then:
1. Think about what exactly you have been asked to do
2. Come up with a number of dishes that would suit the task
3. Decide on one dish (solution)
4. Complete the paperwork for the exam:
   - write out why you picked the dish(es) you did, and list other dishes (at least one other) that you considered
   - write out what you are going to do in preparation
   - list the ingredients you need
   - list the equipment you will need
   - write out a time plan listing what you are going to do first, second and so on
   - evaluation: leave space for this

On the day of the exam you will have half an hour preparation time. During this time you can:
- collect equipment
- weigh ingredients
- wash vegetables
- light gas oven but not set the temperature
- check equipment e.g. electric mixer/food processor.

The examiner will check your preparation. The exam itself will last one and a half hours. During this time you must:
- make dish
- serve dish (examiner tastes it)
- wash up (examiner checks dishes)
- tidy away
- do your evaluation.

THE PRACTICAL COOKERY EXAM

IMPORTANT SKILLS FOR THE PRACTICAL EXAM

Chopping onions

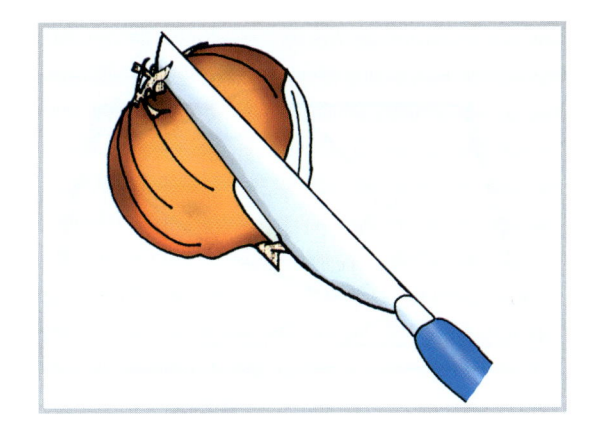

Cut the onion in half from top to bottom.

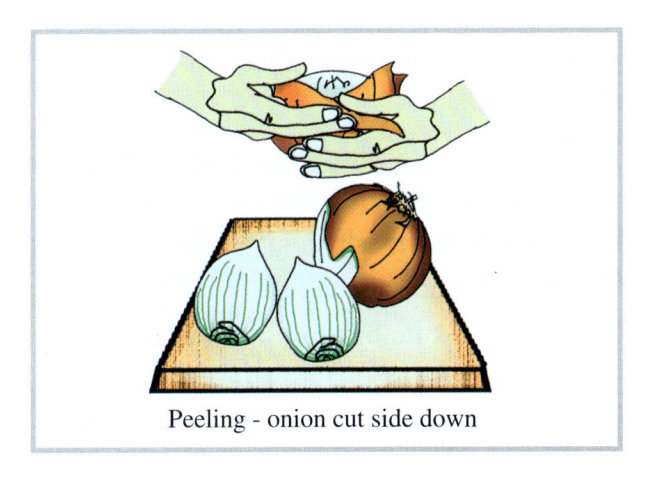

Peeling - onion cut side down

Keeping the cut sides down on the chopping board, remove the skin.

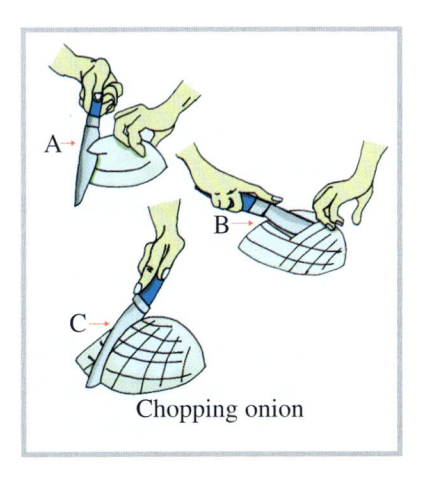

Chopping onion

Make cuts into the onion with a sharp knife as shown in diagrams A, B, C. Keep the cut side down and the root on until the end. Put chopped onion into a glass bowl and cover with a plate until it is used.

Note: Keeping the cut side down on the chopping board helps prevent your eyes watering. Keeping the root on while chopping helps stop the onion falling apart before it is fully chopped.

How to de-seed and chop a pepper

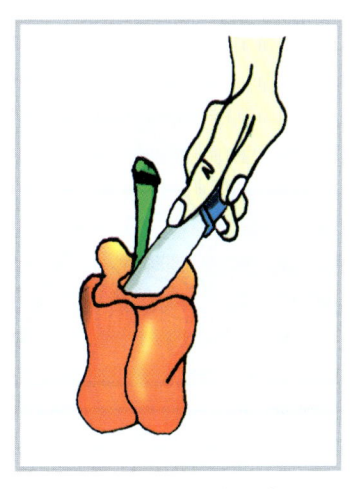

Using a sharp vegetable knife, cut deep into the pepper around the stalk.

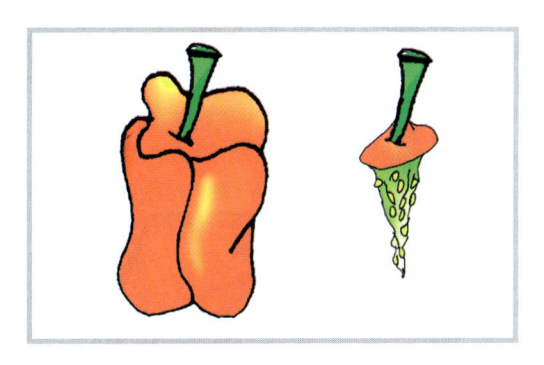

Remove the stalk and seeds.

ESSENTIALS FOR LIVING

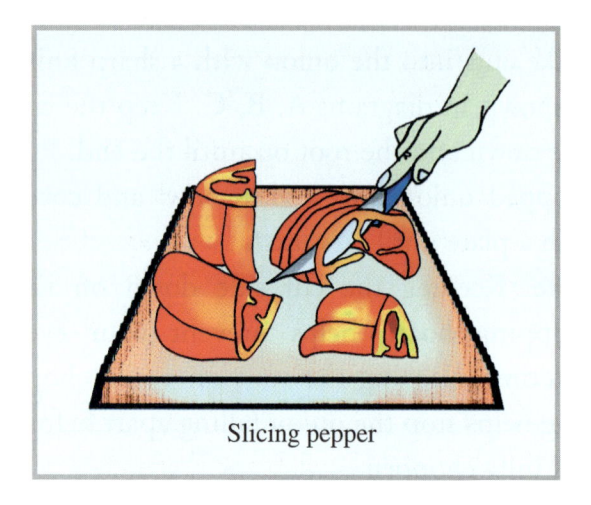

Slicing pepper

Cut the pepper into four pieces. Slice each piece into strips.

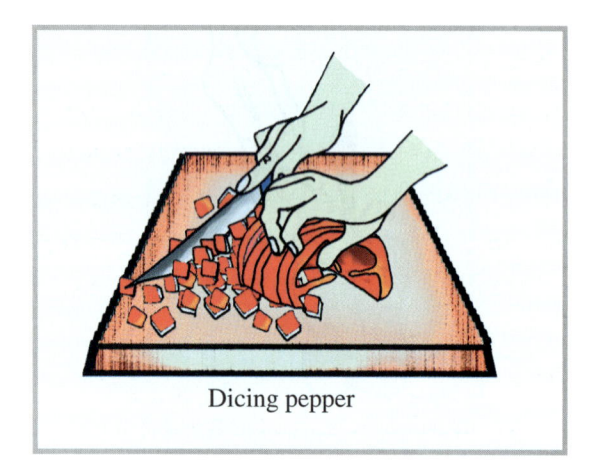

Dicing pepper

Bundle strips together and chop again if wished.

How to chop carrots and parsnips

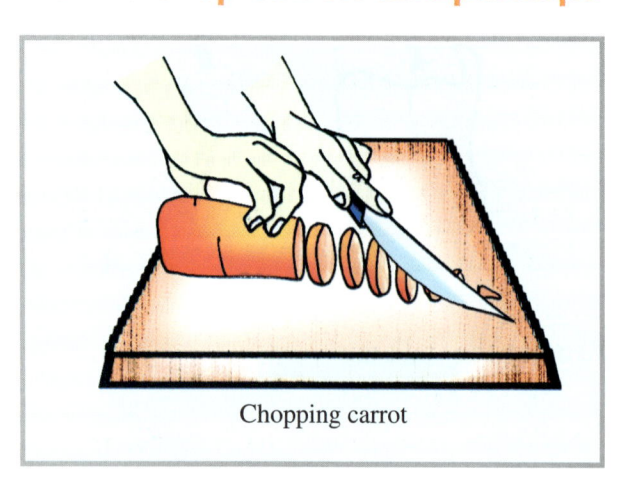

Chopping carrot

Keep the tip of the knife on the chopping board at all times; move the carrot along, not the knife; curl fingers inwards.

How to skin flat fish

Skinning fish

A fish knife is best for this job as it has a flexible blade.
- Lay fish on a chopping board, skin side down, tail towards you.
- Dip fingers in salt to give grip.
- Hold the tail of the fish.
- Slip fish knife between skin and flesh.

Rolling pastry
- Do not make pastry too wet.
- Do not handle pastry too much.
- *Lightly* flour the board and rolling pin.
- Roll in one direction only, e.g. away from you.
- Turn the pastry after every few rolls to stop it sticking.
- Do not stretch pastry: it will just shrink again during cooking.

Washing up
- Use hot water with a good quality washing up liquid.
- Scrape dishes off completely before washing.
- Stack all the dirty dishes to one side.
- Wash using a dishmop or cloth.
- Drain dishes on other side of the sink (on a tea towel if there is no double draining board).
- Fill the sink with cold water.
- Rinse all the dishes.
- Dry thoroughly.

Top tips for the practical exam
When you have read each tip, tick the box with pencil.

1. Write your exam number:
   i) on all your written work
   ii) on a sticker on your apron
   iii) on a large A4 sheet of paper taped to your table

Exam number:	357146
Task Number:	4
Dish:	Spaghetti bolognese
	Garlic bread

2. Make sure you wash your hands, cover cuts, tie back long hair, remove jewellery and have a clean apron on, or else marks will be taken away under 'hygiene'.
3. Take care doing written work for the cookery exam: use the sample below as a guide. Written work makes up thirty per cent of your marks; most of these marks you could have before you walk in the door. Use the form provided in your workbook page 59 (this form may be photocopied).
4. Do not waste anything: time, ingredients, water, gas, electricity.

Time: Know exactly what you have to do so you can work quickly and efficiently. If you have to constantly look at your time plan to see what has to be done next, you are wasting valuable time and may not get finished. (Marks will be taken off for this.)

Ingredients: When peeling vegetables use a peeler; do not take large chunks of skin off with a knife. Try not to have more ingredients than you need. If you have extra do not throw them out; cover them and place in the fridge.

Water: Never leave a tap running. Do not wash anything under a running tap.

Electricity/gas: Turn off cooker rings or oven immediately you finish using them.

5. *Serving food*
   - clear the table before serving food. Serve all the food, not just a portion.
   - garnish or decorate attractively.
   - wipe any spills off serving dishes with kitchen paper.
   - have clean cutlery, a side plate and a bowl of hot water ready for the examiner to taste your food.
   - call the examiner immediately food is served.
   - taste a little yourself so you will be

ESSENTIALS FOR LIVING

able to evaluate. ☐

6. Wash and dry dishes very carefully. Marks are taken off for badly washed dishes. Do not put your dishes away until the examiner has seen them. ☐

7. Do not throw out the contents of the bin until you have shown it to the examiner. ☐

8. Evaluate both the dish and how you think you did in the exam (see sample evaluation). ☐

CASE STUDY

Linda O'Brien has just finished her cookery exam. She had to make a dish for a festive occasion. She picked Christmas and made mince pies. They turned out OK, the pastry was lovely but the mince leaked out of them in the oven so they look a bit burned. Linda knew her recipe very well and was finished fifteen minutes early. Below is her written work. Read through it; it will help you do your own written work.

Sample planning, preparation and evaluation sheet

Exam number: 5831563
Task Number: 4

(4) Many public holidays and festive occasions are traditionally celebrated with special foods or dishes. Name one such event. List the dishes associated with this event. Prepare and cook one of the dishes listed and serve, where appropriate, with a suitable accompaniment.

Analysis: what have you been asked to do?

Write out the task in your own words

I have been asked to pick a festive occasion and to name foods or dishes associated with it. I must then pick one of the dishes I have named and prepare and cook it. If there is something that is usually served with the food/dish I must make it as well (an accompaniment).

THE PRACTICAL COOKERY EXAM

Investigation: List *three* things that you must consider before deciding on what to cook.
1. The dish/food I pick must be traditionally linked to the festive occasion I have named.
2. I must be able to make it in one hour (other half hour for serving, washing up and evaluation).
3. I must think about what is usually served with this dish (accompaniment).

List solutions and ideas: *Two* solutions for each part of the task.
The festive occasion I have chosen is Christmas. Dishes/foods traditional at Christmas are mince pies with cream or brandy cream; Yule log with a fruit coulis.

Decision: Choose *one* solution and fill it in on the menu card below.

MENU CARD

Mince pies and fresh whipped cream

Reasons for your choice of dish(es): Give *two* good reasons.
1. I have chosen to make mince pies accompanied with whipped cream because they are traditional at Christmas and can be made in under one hour.
2. I have chosen mince pies and cream because I have made short-crust pastry many times before and mince pies twice before. I feel confident I could carry out this task well.

Planning and preparation: List all the ingredients and equipment you will need for the task. As you collect them during the half-hour preparation, tick them off on the list.

List of ingredients
Short-crust pastry
  200 g plain flour
  100 g margarine
  pinch of salt
  cold water
Filling
  150 g mincemeat
  egg to glaze
Accompaniment
  250 ml fresh cream
  25 g castor sugar

ESSENTIALS FOR LIVING

List of equipment

mixing bowl
small glass bowl
knife
teaspoon
fork
pastry brush
patty tin
sieve
flour dredger
oven gloves
wire tray
round plate
doily
rolling pin
pastry cutters
electric mixer for cream
dish for cream

Preparation: List everything you are going to do during the half hour preparation time before the exam.
1. I will collect and weigh my ingredients.
2. I will collect all the equipment I need and set my cookery table.
3. I will check the shelf positions and then light my gas oven but not set the temperature on it.
4. I will check that my electric mixer is working properly.
5. I will grease the patty tin lightly.

Time plan: List what you have to do during the exam in the correct order. Don't forget evaluation time at the end.

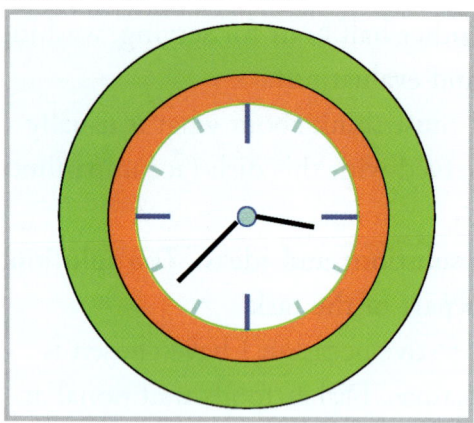

Step 1: Make pastry: sieve flour and salt, add chopped margarine, rub in. Add spoons of water until pastry comes together.
Step 2: Set gas oven to gas mark 6.
Step 3: Divide pastry into 2/3 and 1/3. Roll out big piece. Cut out 12 circles (large cutter), put in patty tin. Roll out rest of pastry, cut out lids (smaller cutter). Put spoon of mince into each. Wet edges. Put lids on, seal with fingers. Pierce with fork, glaze with egg.
Step 4: Bake for 10 minutes at gas 6. Reduce to gas 5, bake for 10 minutes more, until golden. Cool on a wire tray.
Step 5: While pies are baking, clear and wash table, whip cream.
Step 6: Sprinkle pies with icing sugar.

through a sieve. Serve pies on a round plate with a doily. Serve cream in a nice dish.

Step 7: Put out cutlery, plate and bowl of hot water. Call examiner to taste pies.

Step 8: Do wash-up. Show wash-up and bin to examiner. Put dishes away.

Step 9: Do evaluation.

Evaluation: This is not a description of the dish

Colour and presentation: How did you try to make the dish look well? Did you succeed?

The colour of my mince pies was golden brown. However because of the mince leaking out in the oven they were a bit black around the edge which is not good. I tried to make them look good by sprinkling them with icing sugar and putting them on a round plate with a doily.

Texture and taste: Was there more than one texture (contrast)? Was the flavour strong enough/too strong? Would you add anything to improve the texture/flavour?

There was good contrast in texture between the crisp pastry and the smooth cream. I thought the pies tasted really good – not too sweet. You could not taste the burned edges. Maybe next time I could add brandy essence to the cream for more flavour.

Cooked/doneness: Was your dish overcooked, undercooked or just right? how did you check if it was cooked properly?

My mince pies were cooked properly. They looked burned, though, because of the edges. I knew they were cooked because the pastry was golden brown and they were in for the correct amount of time.

Did you meet the brief? Did you do everything you were told to do for your task? Explain why you think you did/didn't.

I think I did meet the brief because mince pies are traditional at Christmas and whipped cream goes well with them.

Your overall performance: What did you do well/not so well?

I made the pastry well. It rolled out easily for me and did not stick to the table. I worked quickly and efficiently. I was finished in plenty of time. Overall I think I was well organised both in making the pies and doing the wash-up. I did, however, put too much mince in the pies.

Are there any changes you would make?

Yes: I would not have put as much mince in the pies and I would have made brandy cream instead of ordinary whipped cream for more flavour.

chapter 10

The food groups

There are *four* groups:
- protein group
- milk and milk products group
- cereal and potato group
- fruit and vegetable group

Questions on the food groups are frequently asked in section B, question 1 (both levels) in the Junior Certificate examination.

PROTEIN GROUP
Meat, fish, eggs, meat alternatives and nuts
Eat two servings per day from this group.

Meat
The term *meat* means the edible flesh and often internal organs of cattle, pigs, sheep, poultry (chicken and turkey) and game (wild animals and birds).

Red meat: cattle, pigs, sheep
Nutritive value: Red meat, while a great source of nutrients, especially protein, iron and vitamin B, should *not* be eaten every day. This is because it contains large amounts of saturated fat, which is linked to heart disease.

Types of red meat:

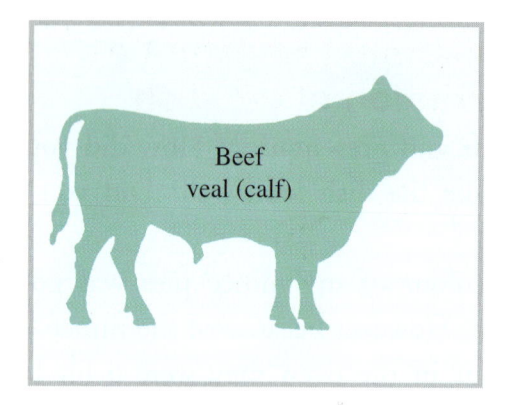

Beef
veal (calf)

Cattle

Pork, bacon, ham, products e.g. sausages

Pigs

Mutton (sheep), lamb, lamb's liver and kidney are often eaten

Sheep

66

THE FOOD GROUPS **10**

Activity 10.1 – Workbook p. 64

Composition: What nutrients are in red meat?

Higher level

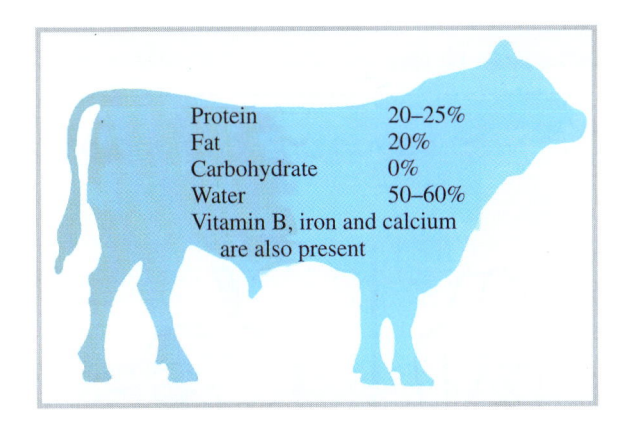

Protein   20–25%
Fat   20%
Carbohydrate   0%
Water   50–60%
Vitamin B, iron and calcium are also present

Activity 10.2 – Workbook p. 64

Under the microscope: The structure of meat

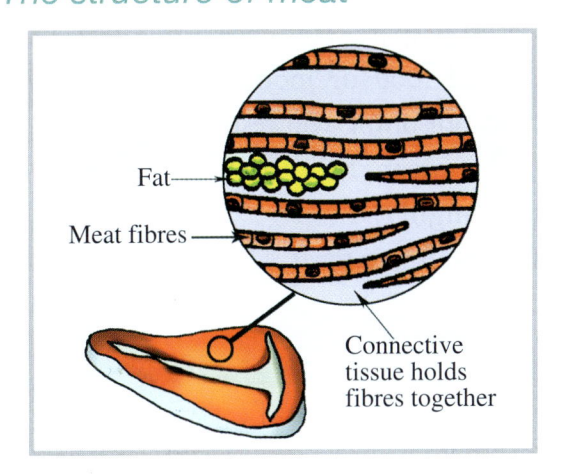

The structure of meat

Tough meat has more tough connective tissue and comes from old animals or very active parts of the animal, such as the leg or neck.

Tender meat has less connective tissue and comes from young animals or parts of the animal that do not move much.

Five ways of making meat tender

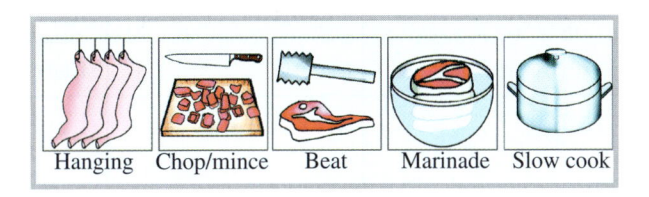

Hanging   Chop/mince   Beat   Marinade   Slow cook

- hang for a few days at the butcher's
- chop or mince
- beat with a steak hammer
- marinade – soak in flavoured liquid
- slow moist cook e.g. stew

Cuts of meat

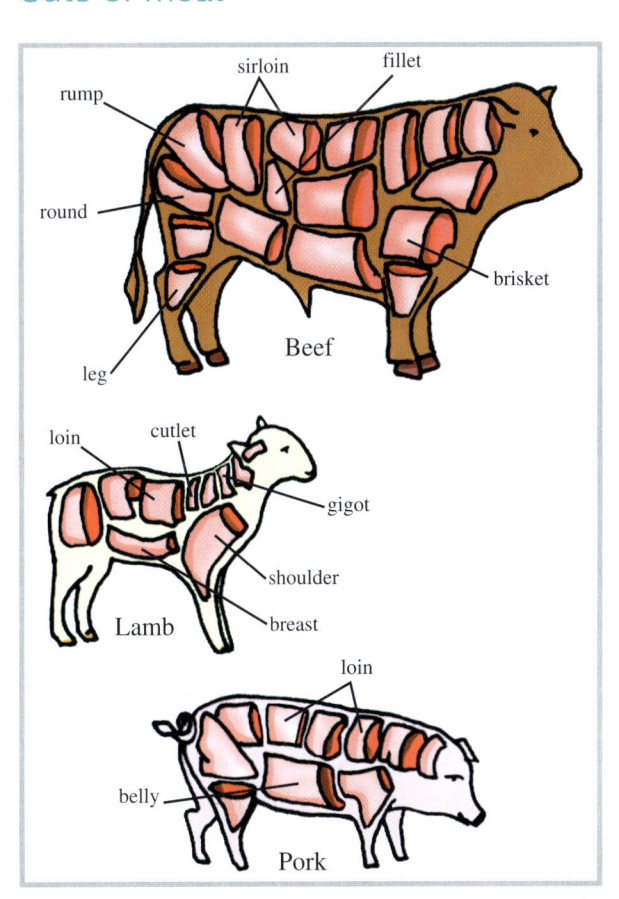

Beef: sirloin, fillet, rump, round, brisket, leg

Lamb: loin, cutlet, gigot, shoulder, breast

Pork: loin, belly

Note: Offal, which are internal organs such as liver and kidney, are cheap, nutritious and cook quickly, for example by grilling.

67

Type of cut	Beef	Lamb	Pork	How to cook
Tender	sirloin	fillet, loin, cutlets	loin	grill, fry or roast
Medium	round	shoulder	–	slow roast
Tough	neck, chuck, leg	shank, gigot	shank, shoulder, gigot	stew, mince
Fatty	brisket	breast	belly	stew
The more tender the cut the more expensive it is.				

When buying meat:
- buy from a clean, reliable shop
- meat should have no bad smell
- don't buy fatty pieces
- check both sides of the meat; if pre-packed you are taking a chance
- buy a cut suited to your cooking method, for example there is no point in frying tough gigot chops

Remember: Tough cuts are just as nutritious and as tasty as expensive cuts if cooked correctly.

Activity 10.3 – Workbook p. 65

Safe storage of meat
- meat is a high protein food – bacteria like it – use within two days
- check best-before dates
- store in fridge on a large plate so it cannot drip; cover loosely with a clean tea towel

Safe preparation and cooking of meat

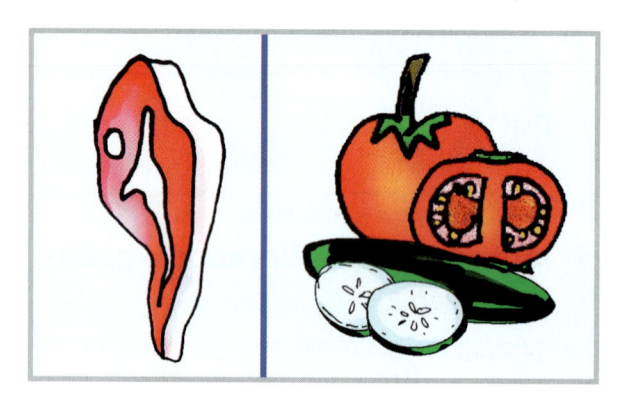

- never prepare raw meat and fish near cooked food or food that is eaten raw; food poisoning will result
- thaw meat fully before cooking it thoroughly
- never refreeze meat

THE FOOD GROUPS

Effects of cooking on meat

Effects of cooking on meat

- fat melts
- meat shrinks
- colour changes from red to brown
- protein coagulates, surface seals
- flavours develop
- bacteria are destroyed

Exam time – Workbook p. 65 – Red meat

Poultry: chicken, turkey, duck, goose

Chicken

Chicken is an excellent food: it is high in protein yet low in saturated fat. It is, however, lower in iron than red meat but can be served with iron-rich foods such as green vegetables.

Activity 10.4 – Workbook p. 66

Buying fresh chicken	Buying frozen chicken
• clean shop • check expiry date • no unpleasant smell • flesh firm and plump • flesh not discoloured	• chicken should be below the 'load line' in shop's cabinet • packaging not torn • frozen solid • not discoloured in any way
Storing fresh and frozen chicken	Roasting a chicken
Fresh • remove wrapper and giblets • place on a large plate, cover loosely with a clean tea towel and refrigerate • use within 3 days **Frozen** • place in home freezer as soon as possible after buying it	• defrost completely • preheat oven to 200°C / Fan 190°C / Gas 6 • remove giblets, wash under cold tap, pat dry with kitchen roll • season inside and out • cook 20 minutes per 500 g + 20 minutes extra

ESSENTIALS FOR LIVING

Activity 10.5 – Workbook p. 67

> *Food safety*
> To reduce the dangers of food poisoning from harmful salmonella bacteria follow these simple rules:
> - always thaw frozen, whole chicken properly: twenty-four hours in the fridge or eight hours at room temperature is necessary to defrost it properly
> - always make sure chicken is cooked right through to the bone; when the thickest part of the flesh is pierced with a skewer, the juices should run clear, not pink
> - after handling chicken, wash your hands and utensils in hot soapy water
> - prepare and store raw chicken separately from cooked foods or foods eaten raw

Meat alternatives

Tofu

Tofu is a white, creamy, high-protein food made by separating soya milk into curds (tofu) and whey and then pressing the curds into cubes or blocks. Tofu can be used in the following ways:
- tofu can be used much like cheese in vegan pizza
- thread cubes of firm tofu onto skewers with mushrooms, cherry tomatoes, and onions. Marinade in soy sauce, brush with olive oil and grill to make tasty kebabs (see photo)
- whizz tofu with strawberries and honey to make a dairy-free fruit fool
- mix with onion, garlic and herbs; form into burgers; coat in flour or egg and breadcrumbs; grill or fry
- cream tofu and add flavourings, e.g. garlic, to make tasty nutritious dips

Tofu kebabs and TVP burgers

TVP (Textured Vegetable Protein)

This protein-rich food is made from soya beans. It is flavoured and shaped to resemble meat: chunks, mince and steaks. TVP is usually bought dried and once reconstituted with water can be used in a huge variety of dishes. For example lasagne, spaghetti bolognese, burgers (see photo), stew, shepherd's pie etc. Some non-vegetarians use one-third TVP, two-thirds meat in dishes because it is healthier. 'Quorn' is one well-known brand of TVP.

Pulse vegetables (peas, beans and lentils) and nuts are also used as meat substitutes because of their high protein content. Pulses often need soaking before use.

Fish

As a nation, Irish people tend not to eat as much fish as other island nations. We should eat more fish as it is an easily cooked, nutritious alternative to red meat.

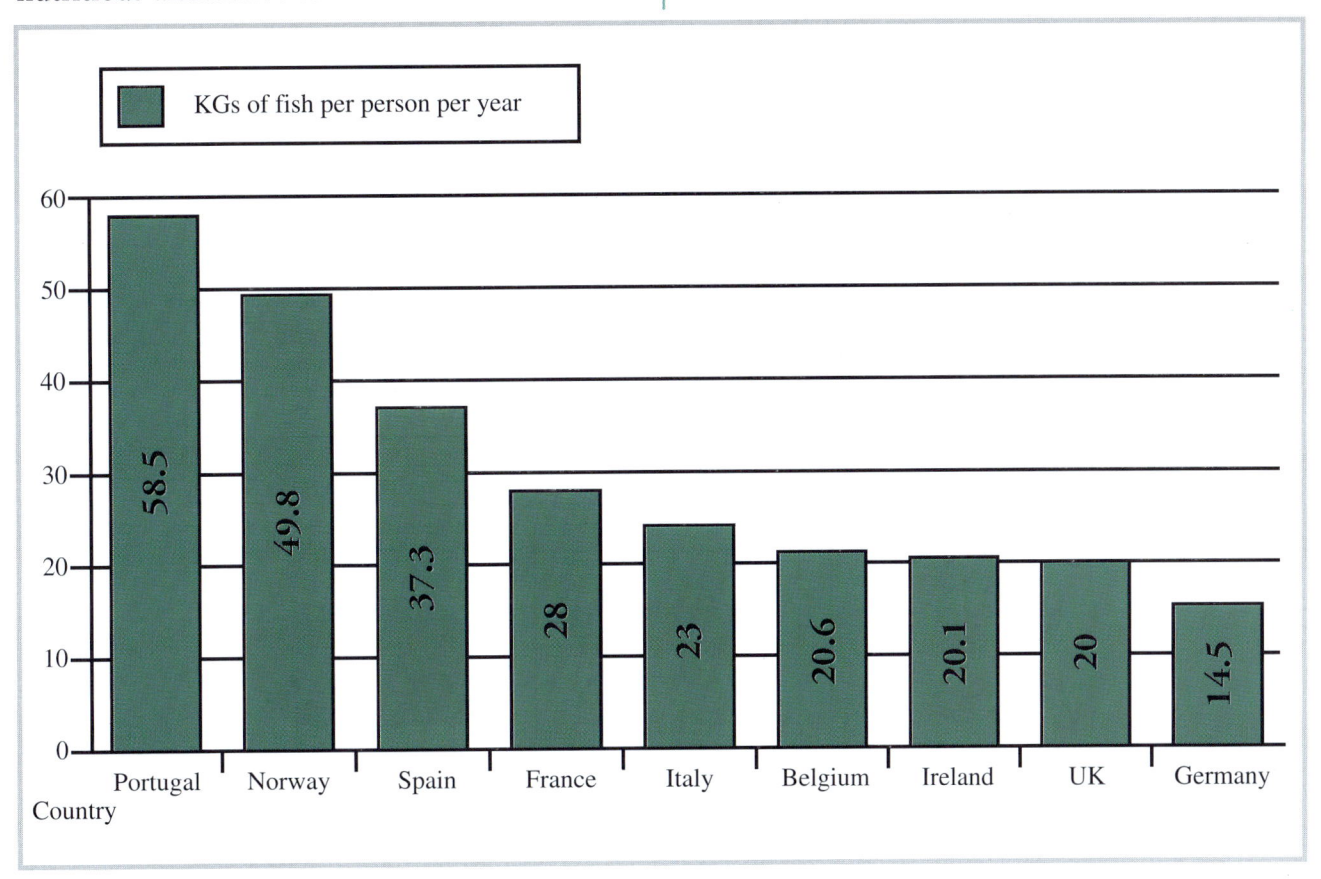

Country	KGs of fish per person per year
Portugal	58.5
Norway	49.8
Spain	37.3
France	28
Italy	23
Belgium	20.6
Ireland	20.1
UK	20
Germany	14.5

Fish consumption in Europe

Classification of fish

Fish can be classified in three ways:

1. **Shape:**
   Round, e.g. herring
   Flat, e.g. plaice
2. **Habitat (where they live):**
   Salt water, e.g. mackerel
   Freshwater, e.g. brown trout
3. **Nutritive value:**
   White fish, e.g. plaice, sole, cod, haddock, whiting
   Oily fish, e.g. mackerel, salmon, herring, trout, sardines
   Shellfish, e.g. crab, prawns, lobster, mussels

Unlike oily fish, which contains up to twenty per cent fat (mostly unsaturated fat), white fish contains no fat: the fat of a white fish is stored in its liver, which is removed before cooking. Shellfish contain some fat (mostly unsaturated).

ESSENTIALS FOR LIVING

Plaice	Cod	Haddock
Mackerel	Salmon	Herring
Crab	Prawns	Mussels

✏️ **Activity 10.6 – Workbook p. 67**

✏️ **Activity 10.7 – Workbook p. 68**

Nutritive value

- Fish is an excellent source of protein. Unlike red meat it does not contain saturated fat and so will not raise cholesterol levels.
- White fish contains no fat and is useful in the diets of slimmers. All fish is a good source of vitamin B and oily fish also contains vitamins A and D.
- All fish contains the mineral iodine, canned fish also contains calcium if the bones are eaten.
- Fish contains no carbohydrate and may be served with a carbohydrate-rich food such as potatoes. Fish also lacks vitamin C, and should therefore be served with a wedge of lemon.

✏️ **Activity 10.8 – Workbook p. 68**

✏️ **Activity 10.9 – Workbook p. 68**

THE FOOD GROUPS   10

Buying fish

Fresh	Frozen
Buy fish in season from a clean shop. Fish must be very fresh: • moist unbroken scales • firm flesh, clear skin and markings • no bad smell or discolouration • eyes bulging • gills red and not sticky	• frozen solid • stored below 'load line' in the freezer • wrappings untorn • well within expiry date • put fish into home freezer as soon as possible after buying it

How fish is sold

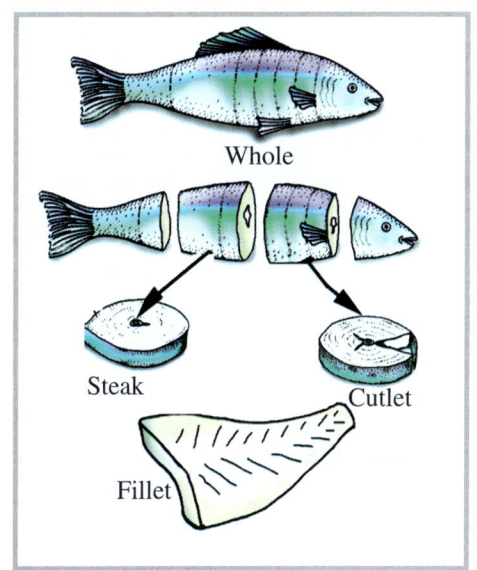

Whole

Steak   Cutlet

Fillet

■ whole (small fish)
■ fillet
■ cutlet
■ steak: large round fish sliced into cutlets (from the head end) or steaks (tail end)

Activity 10.10 – Workbook p. 69

Preparing fish

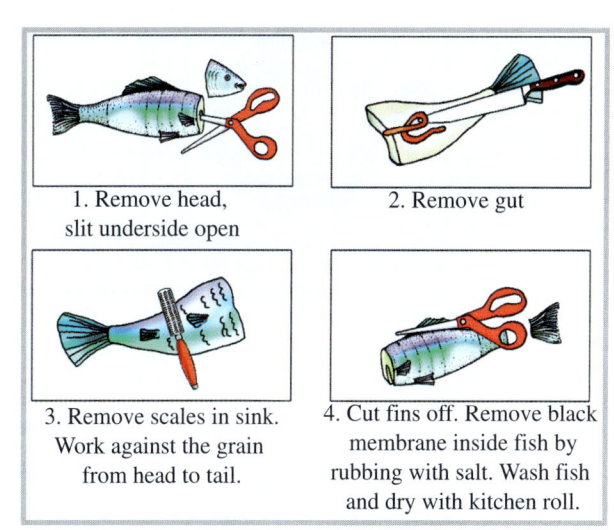

1. Remove head, slit underside open
2. Remove gut
3. Remove scales in sink. Work against the grain from head to tail.
4. Cut fins off. Remove black membrane inside fish by rubbing with salt. Wash fish and dry with kitchen roll.

Preparing round fish

Round fish prepared like this can be stuffed and oven-baked. Round fish can also be filleted: two fillets are removed by paring them away from head to tail; keep the knife as close as possible to the ribcage.

ESSENTIALS FOR LIVING

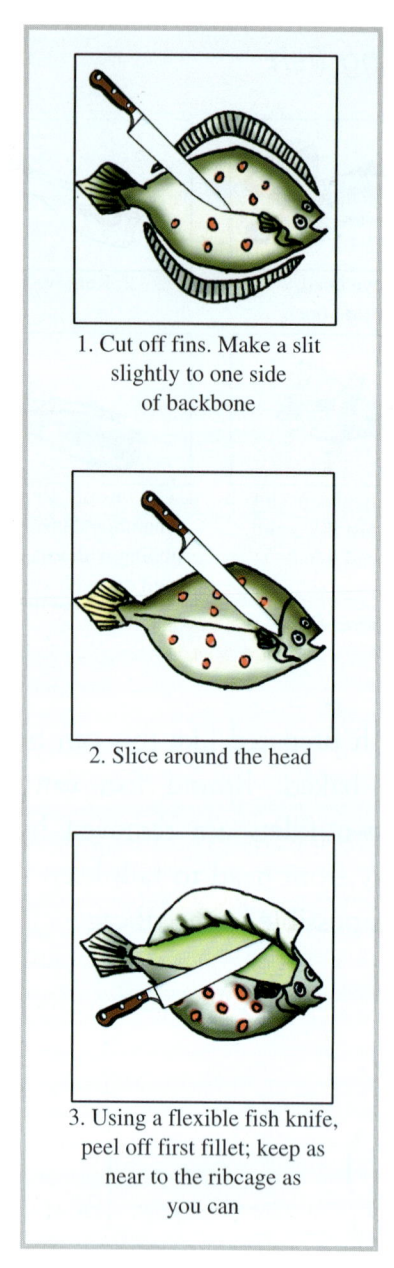

1. Cut off fins. Make a slit slightly to one side of backbone

2. Slice around the head

3. Using a flexible fish knife, peel off first fillet; keep as near to the ribcage as you can

Flat fish – filleting

How to skin fish

Skinning a fish fillet

A special flexible fish knife makes this easier.

1. Place fish skin side down, tail towards you on a chopping board.
2. Sprinkle salt on tail end to give you grip.
3. Slide the knife between skin and flesh; keep the knife as close to the skin as possible.

Cooking fish

Fish cooks very quickly; if overcooked it breaks up. It loses its translucent (see-through) appearance, turning white. Micro-organisms are destroyed and some vitamins and minerals may dissolve into the cooking liquid.

- Protein coagulates (sets) + shrinks
- Fish becomes opaque
- Connective tissue dissolves + fish breaks apart easily
- Microorganisms are killed
- Minerals + Vits dissolve into cooking liquid
- Heat destroys some vit B.

milk - poaching

Methods of cooking fish

1. Baked (may be stuffed or baked in a sauce)

2. Steamed (in a steamer or between two plates)

6. Stewed (firm fish only, e.g. monkfish)

3. Poached (make sure liquid is not boiling)

5. Fried (fish often coated)

4. Grilled (very quick method, nutrients retained)

Once cooked, fish is usually garnished with a sauce (such as tartar sauce), parsley, and a wedge of lemon (see sauces, page 106 and garnishes, page 36).

Processed fish

Smoked fish

Smoked
- example: smoked haddock fillets
- smoking preserves fish

Frozen
- nutrients retained
- expensive
- no waste
- coated fish and ready dinners available

Canned/bottled
- in oil or brine
- tuna, salmon, sardines and shellfish are often canned or bottled

ESSENTIALS FOR LIVING

Exam time – Workbook p. 69 – Fish

Eggs

Uses of eggs in cookery

Nutritive value of eggs

Eggs lack carbohydrate and so are often served with carbohydrate-rich foods, for example boiled egg and toast.

Eggs lack vitamin C and are often served with foods rich in vitamin C, for example omelette served with a side salad.

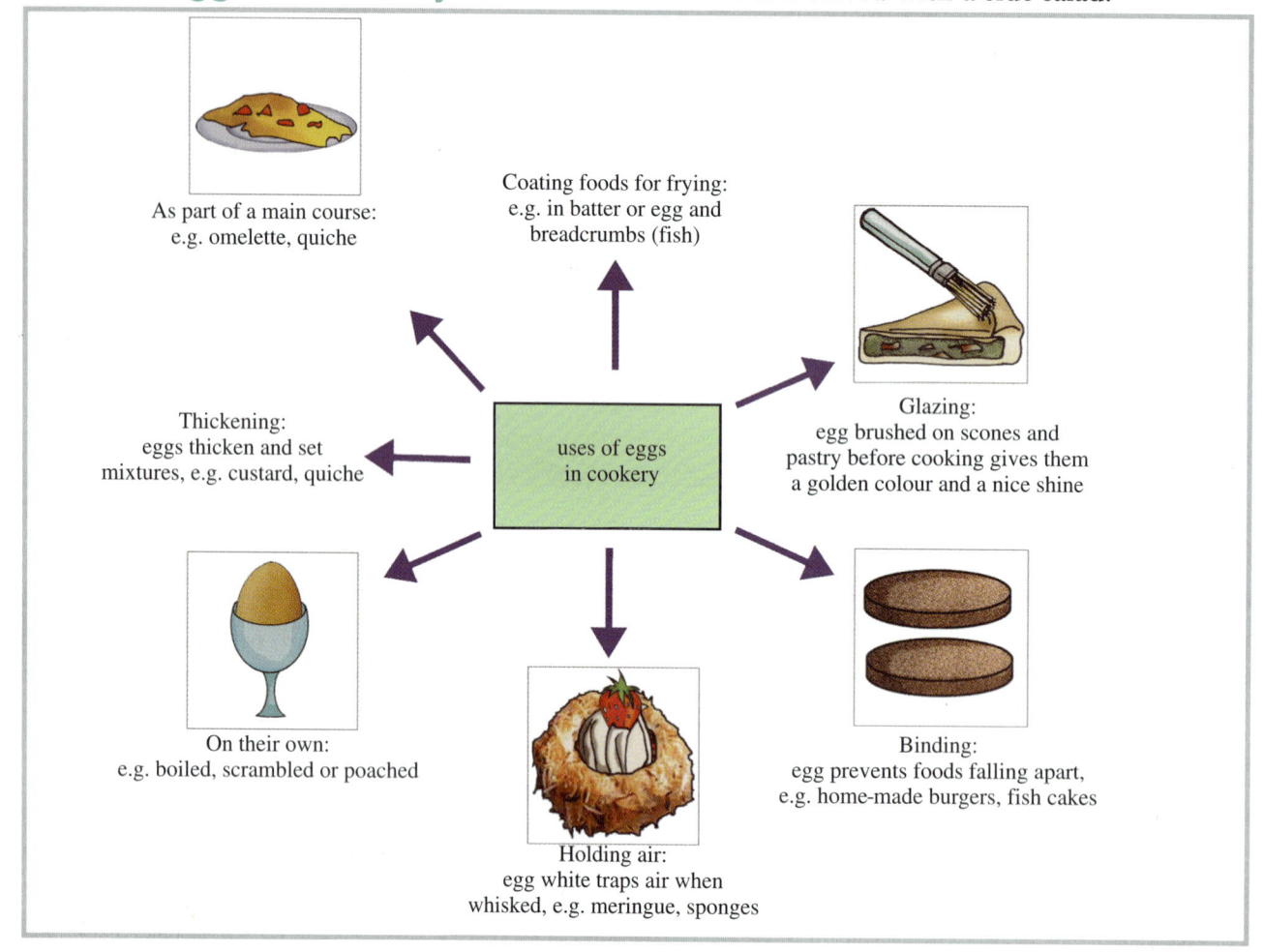

uses of eggs in cookery

- As part of a main course: e.g. omelette, quiche
- Coating foods for frying: e.g. in batter or egg and breadcrumbs (fish)
- Glazing: egg brushed on scones and pastry before cooking gives them a golden colour and a nice shine
- Thickening: eggs thicken and set mixtures, e.g. custard, quiche
- On their own: e.g. boiled, scrambled or poached
- Holding air: egg white traps air when whisked, e.g. meringue, sponges
- Binding: egg prevents foods falling apart, e.g. home-made burgers, fish cakes

Average composition of eggs
Higher level

Protein	Fat	Carbohydrate	Minerals	Vitamins	Water	Cholesterol
13%	11%	0	iron, calcium 1%	A, B, D	75%	385 mg / 100g

Activity 10.11 – Workbook p. 71

Activity 10.12 – Workbook p. 71

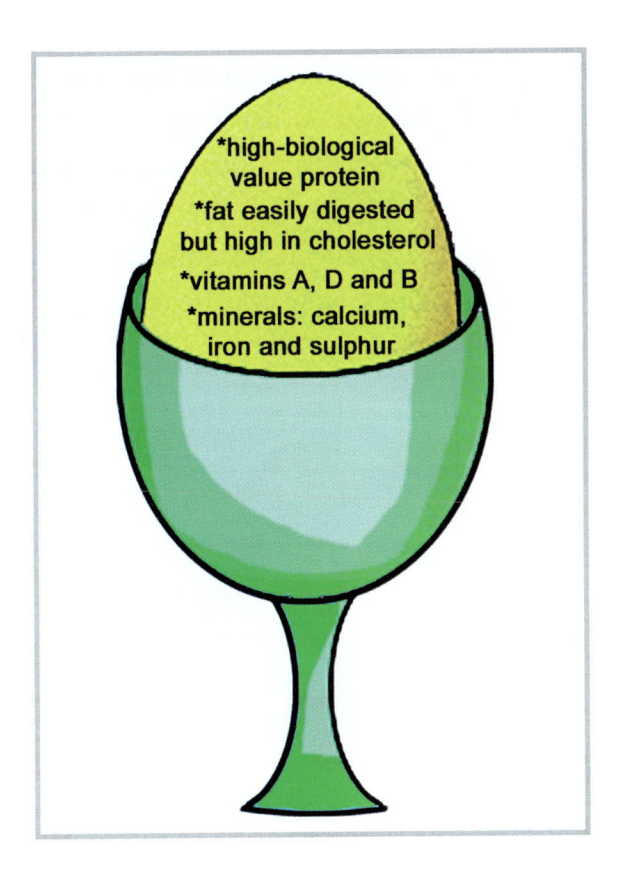

- *high-biological value protein
- *fat easily digested but high in cholesterol
- *vitamins A, D and B
- *minerals: calcium, iron and sulphur

Buying eggs

All the information you need to buy good quality, fresh eggs must be written on the box they come in. Check that all eggs in the box are unbroken.

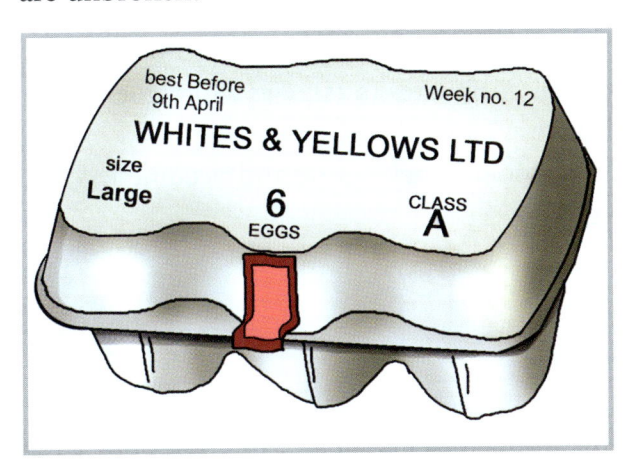

1. **Name of the producer:** must be on the box.
2. **Class:** eggs are graded A, B, C or Extra. Extra is the best, followed by A.
3. **Size:** eggs are now graded in four sizes; extra large (XL), large, medium and small. Use large for cooking.
4. **Country of origin**
5. **Expiry or best-before date:** this is usually three weeks from the day the eggs were laid. Eggs sold loose must have this stamped on each one.
6. **Week number:** freshness can easily be judged by looking at this number. Eggs are given a number (1–52), week 1 being the first week in January.

Free-range hens

Battery hens

ESSENTIALS FOR LIVING

Free-range hens' eggs must have 'free range' stamped on the box or on each individual egg if sold loose.

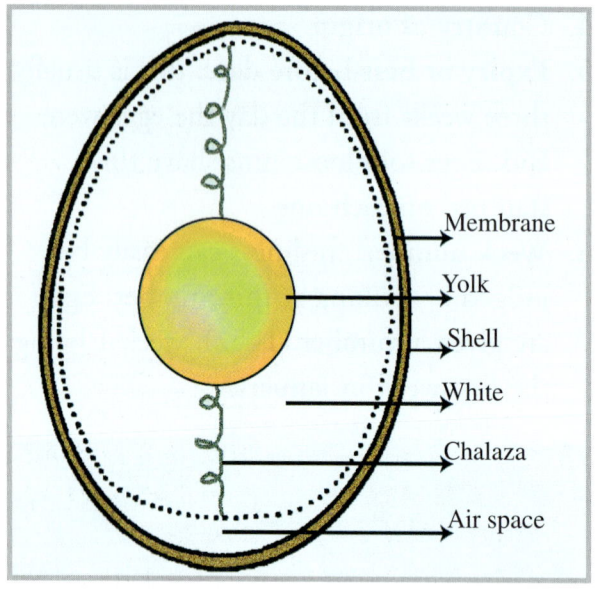

Structure of an egg

- Membrane
- Yolk
- Shell
- White
- Chalaza
- Air space

Checking if an egg is fresh

If eggs have been removed from their box and placed in the egg compartment of the fridge you may have no expiry date to tell you whether they are fresh or not. Eggs can be tested for freshness in other ways.

- As an egg gets stale water evaporates out of the egg through the shell and air passes in. The air sac becomes larger and the egg lighter. Stale eggs float.

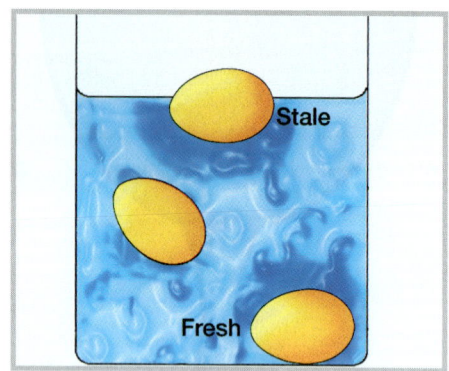

Testing an egg for freshness

- Crack open an egg onto a saucer. It is fresh if the yolk is raised and the white around it is jelly-like. A stale egg has a flat yolk and a runny white.

Storing eggs	Effects of cooking on eggs
• in the fridge, pointed end down • away from strong smelling foods • if storing yolks separately, place in a bowl of water in fridge • store whites in an air-tight container in the fridge	• protein coagulates (sets): egg white goes from clear to white • if overcooked they curdle and become less digestible

Using eggs in cookery
• bring eggs to room temperature before using • egg whites will not whisk properly if there is yolk present or if the bowl or whisk has even a trace of grease on it • to prevent curdling don't overcook eggs: when adding egg to a hot liquid, e.g. milk for egg custard, cool the liquid somewhat and then add it to the egg, not the other way round • raw or lightly cooked eggs should not be given to babies, toddlers, pregnant women, elderly people or invalids due to the risk of salmonella poisoning

THE FOOD GROUPS   10

Activity 10.13 – Workbook p. 71

Exam time – Workbook p. 72 – Eggs

Revision crossword – Workbook p. 73

MILK AND MILK PRODUCTS

Milk, butter, cream, yogurt and cheese
Eat three or four servings per day from this group.

Milk

Food value

Milk is an almost perfect food. Babies thrive on milk alone for the first three to four months of life. However:

- milk lacks vitamin C
- milk lacks iron

Apart from this milk is very important in the diet as:

- milk is a good source of protein for growth and repair
- the fat in milk is easily digested, and so milky foods are suited to those recovering from illness

- low-fat milks are available for slimmers
- milk is rich in calcium: 1 glass (250 ml) provides an adult with over half their daily calcium
- milk also contains vitamins A, B and D
- cow's milk should never be given to babies under one year as it is too concentrated

Effects of heat on milk	Buying and storing milk
• flavour changes • bacteria destroyed • protein coagulates and skin forms, steam builds up under skin: can boil over • loss of vitamin C and B	• check expiry date • store in fridge away from strong smelling foods e.g. onions • use in order of expiry dates • don't mix milks with different expiry dates

Activity 10.14 – Workbook p. 74

Activity 10.15 – Workbook p. 74

Composition of whole milk

Protein	Fat	Carbohydrate	Minerals	Vitamins	Water
3.5%	4%	4.5%	Calcium	A,B,D	87%

ESSENTIALS FOR LIVING

Types of milk

1. **Whole milk:** ordinary milk; four per cent fat; recommended milk for children

2. **Low-fat milk:** sometimes called semi-skimmed, less than half the fat of whole milk

3. **Skimmed:** all but a trace of fat removed; flavour altered and vitamins A and D also removed; not for children

4. **Fortified (super-milk):** a low-fat milk with vitamins A and D and calcium added

5. **Buttermilk:** acidic, used in baking

6. **Pasteurised:** pasteurisation kills the bacteria in milk to make it safe without spoiling the taste of the milk; milk is heated to 72°C for 15 seconds and then cooled quickly. All milk is pasteurised.

7. **Homogenised:** homogenisation spreads the cream or fat evenly throughout the milk; most milk is treated in this way

Processed milks

UHT (**long-life**): heated to 132°C, lasts unopened for months; available in cartons or small single portions (hotels)

THE FOOD GROUPS 10

Evaporated milk: some water removed, then canned. Long-lasting; used in dessert-making

Dried milk: all water removed; make up by adding water. Taste is altered

Condensed milk: similar to evaporated milk but with sugar added

Soya milk: made from soya beans; suited to vegans and those with dairy allergies

ESSENTIALS FOR LIVING

Activity 10.16 – Workbook p. 75

Exam time – Workbook p. 75 – Milk

Milk products: cream and butter

Please note that while cream and butter are milk products they are not in the milk, yogurt and cheese group. They belong at the top of the food pyramid and should be avoided or eaten sparingly.

Cream

Cream is fat that rises to the top of milk and is then removed. Cream is very high in kilocalories.

Types of cream

Types of cream

Type	Description
Standard cream	carton often states 'fresh cream'; 40% fat
Double cream	used in dessert making; 48% fat
Low-fat cream	reduced fat: still contains 30%
Soured cream / crème fraîche	cream treated with a lactic acid culture: this pleasantly sours and thickens it; used for dips, dressings etc.
Aerosol	keeps longer, quick and convenient

Butter

When cream is churned the fat comes together as *butter*. The liquid that runs off is called *buttermilk*. Butter is usually salted. Butter is very high in fat – *eighty per cent*. Low-fat or 'light' butter has half this fat content. *Dairy spreads* such as Dairy Gold are a mixture of butter and vegetable oil; they contain roughly the same amount of fat but less cholesterol. *Polyunsaturated margarines* such as Flora are very low in cholesterol and are a much healthier option.

Activity 10.17 – Workbook p. 77

THE FOOD GROUPS

Yogurt

Yogurt is a milk product made by adding souring lactic acid bacteria (harmless) to milk. Yogurt has all the nutrients of milk. Low-fat varieties are available; others sometimes have sugar added.

Types of yogurt

1. **Natural yogurt:** white, unflavoured yogurt; a good base ingredient for savoury dips etc.
2. **Set yogurt:** e.g. petit filous
3. **Yogurt drinks:** milk and flavours added to plain yogurt e.g. Yop
4. **Greek yogurt:** thick, creamy unflavoured yogurt
5. **Fruit yogurt:** either real fruit is added, or just fruit juice if a smooth fruit-flavoured yoghurt is required.

Uses of yogurt

| Healthy alternative to cream | Healthy dessert on its own, great in packed lunches |
| Natural or Greek for salad dressings | Dips: natural or Greek with other ingredients added, e.g. chives |

Cheese

Classification

Cheese is classified into hard, semi-hard, soft or processed types.

Hard cheese – Examples: Cheddar, Parmesan

Semi-hard cheeses – Examples: Edam, Irish blue

ESSENTIALS FOR LIVING

Soft cheese – Examples: cream cheese e.g. Philadelphia and cottage cheese

Also available are processed cheeses such as Easi Singles.

Composition of hard cheese

Higher level

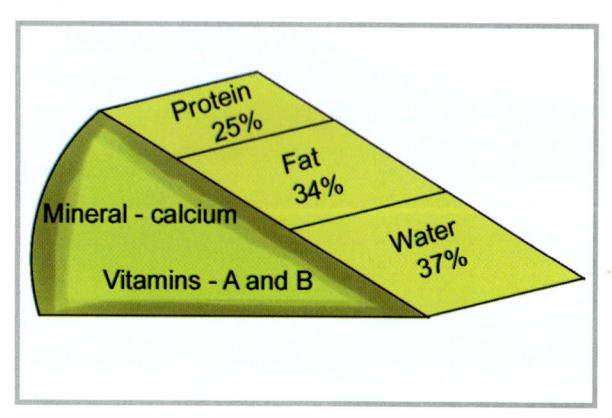

- Protein 25%
- Fat 34%
- Water 37%
- Mineral - calcium
- Vitamins - A and B

Activity 10.18 – Workbook p. 77

Food value

- Cheese is an excellent source of both calcium (healthy bones and teeth) and protein (growth and repair). For this reason cheese is great in the diets of children, teenagers, pregnant and breast-feeding women and vegetarians.
- Cheese is very high in fat and cholesterol, however. It should only be eaten in moderation, especially by those on low-fat diets or with high cholesterol.
- Cheese lacks carbohydrate and is generally served with carbohydrate-rich foods such as bread, pasta or crackers.

Uses of cheese in cookery

Uses of cheese in cookery

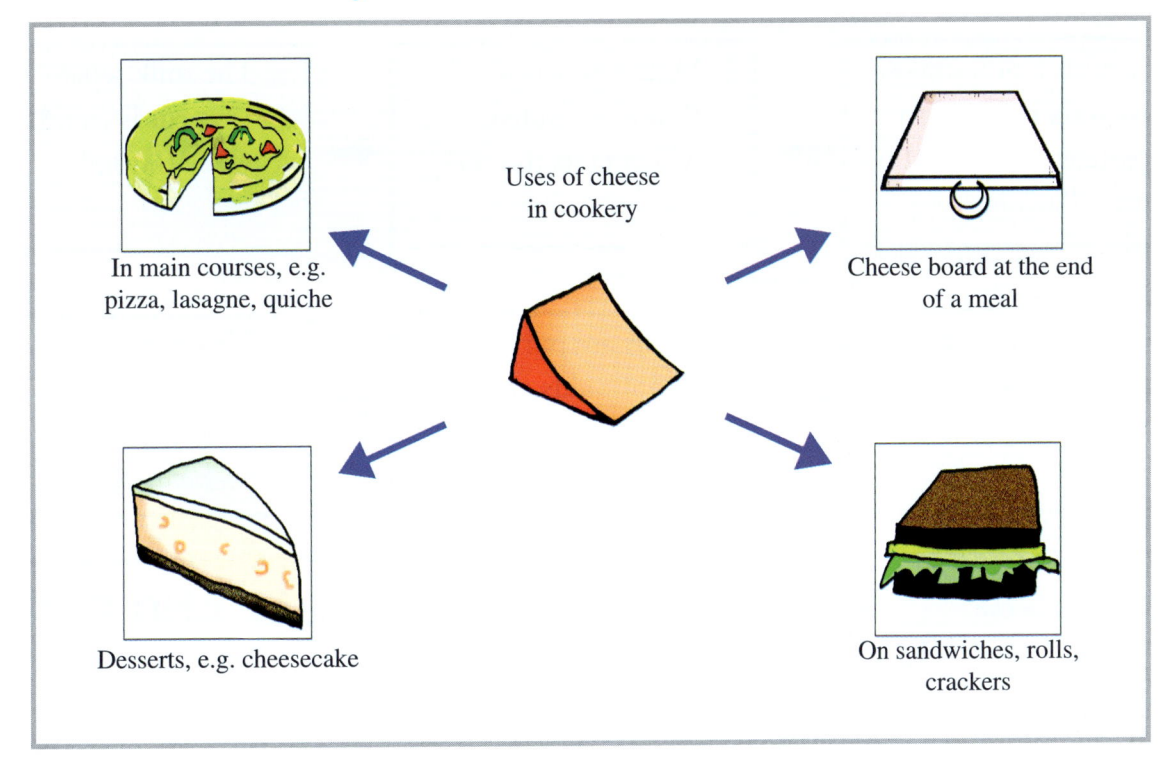

- In main courses, e.g. pizza, lasagne, quiche
- Cheese board at the end of a meal
- Desserts, e.g. cheesecake
- On sandwiches, rolls, crackers

Effects of cooking

- *Eat raw* if possible; cooked cheese is less digestible
- *Protein coagulates* (sets); cheese shrinks
- *Fat melts;* if overcooked fat separates from cheese and cheese goes rubbery, oily and indigestible
- *Mustard* helps make cheese more digestible

The six stages of cheese making

1. A culture of harmless bacteria is added to pasteurised milk, this gives flavour. → 2. Milk is warmed. Rennet is added, which clots the milk. → 3. The milk separates into curds (solid) and whey (liquid).

4. Curds are chopped, pressed and salted. Whey is drained off. → 5. Curds are put into moulds and pressed – firmly for hard cheeses. → 6. Cheese is left to mature. Flavours develop over three to twelve months.

Buying cheese	Storing cheese
• buy in small quantities as it will go off quickly once opened • check expiry date	• cover with grease proof paper, then foil • refrigerate • remove from fridge 1 hour before use to let flavours develop

Activity 10.19 – Workbook p. 77

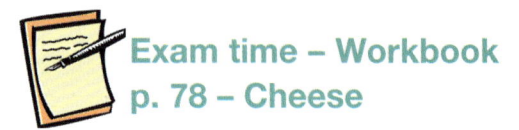

Exam time – Workbook p. 78 – Cheese

THE CEREAL AND POTATO GROUP

Eat six servings per day.

Note: Potatoes are included in the section on vegetables.

Most of the food we eat should come from this food group. All over the world cereals are eaten because they are cheap, nutritious, filling and are easily prepared.

THE FOOD GROUPS

Common cereals and some of their products

Wheat: flour, bread and cakes, pasta, breakfast cereals, e.g. Weetabix, noodles, couscous

Rice: rice (brown rice), Rice Krispies, rice cakes, rice flour, ground rice

Barley: pearl barley, barley water drink, whiskey, beer

Maize: cornflakes, corn on the cob, sweet corn, corn flour, pop corn, corn oil

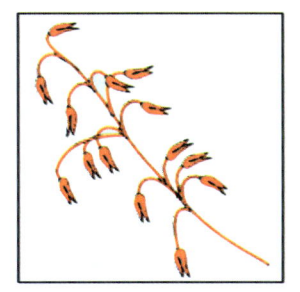

Oats: porridge oats, breakfast cereals e.g. muesli, Readybrek, Common Sense

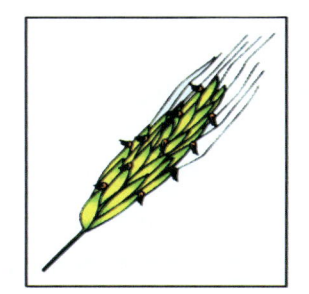

Rye: rye flour, rye bread (dark colour), crispbread e.g. Ryvita, rye whiskey

Activity 10.20 – Workbook p. 80

Average composition of wheat

Higher level

Protein	13%
Fat	2%
Carbohydrate	70%
Minerals	calcium and iron
Vitamins	B group
Water	14%

Structure of cereals

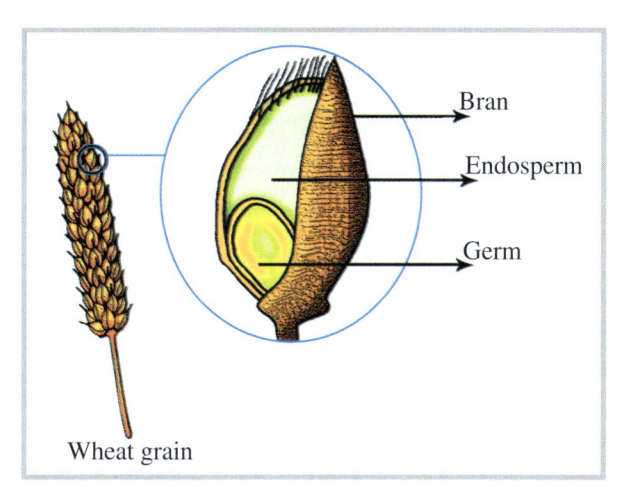

Wheat grain

Bran
Endosperm
Germ

This is a drawing of a wheat grain. Other cereals have a similar structure.

ESSENTIALS FOR LIVING

The bran layer: This part of the grain contains fibre, iron and vitamin B. It is removed during processing, for instance in making white flour.

The endosperm: Starch and gluten. This is the part of the grain that remains after processing.

Germ: This nutritious part of the grain is also removed during processing. It can be bought on its own as wheat germ. It contains protein, fat and vitamin B.

Nutritive value: value in the diet

- *Carbohydrate* is the main nutrient found in cereals. Unprocessed cereals contain fibre that is essential for a healthy digestive system.
- Cereals also contain *protein*. *Gluten* is a protein found in some cereals, such as wheat. While gluten gives bread dough its elastic quality and allows it to rise, people with coeliac disease cannot digest gluten and must avoid eating it.
- Cereals do contain fat, although this is mainly contained in the germ, which is removed during processing.
- Cereals contain the minerals *calcium* and *iron* and some *B group vitamins*. If these are removed during processing they are often put back again. You will see, for example, 'Fortified with calcium and iron' written on such products.

Flour

Although flour can be made from other cereals such as corn, wheat flour is the most common in this country. The type of wheat flour produced depends on the type of wheat used and how much processing it gets.

Processing flour

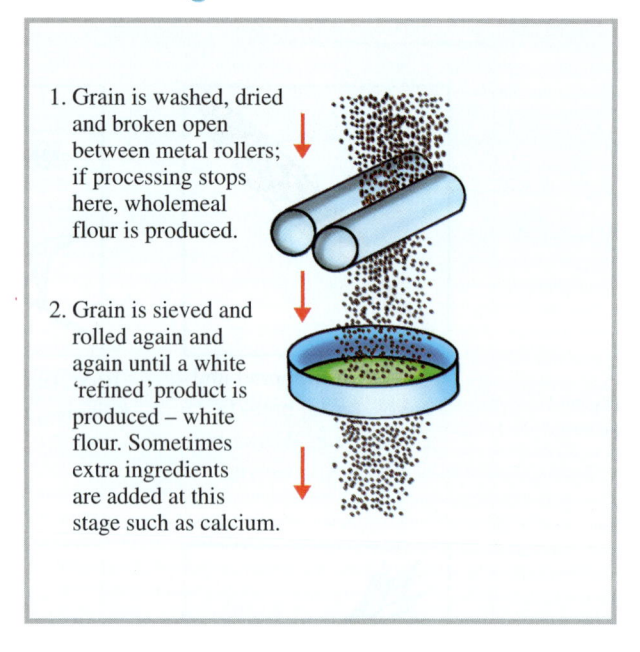

1. Grain is washed, dried and broken open between metal rollers; if processing stops here, wholemeal flour is produced.

2. Grain is sieved and rolled again and again until a white 'refined' product is produced – white flour. Sometimes extra ingredients are added at this stage such as calcium.

A refined cereal product is a processed product from which much of the bran and the germ have been removed.

Types of flour

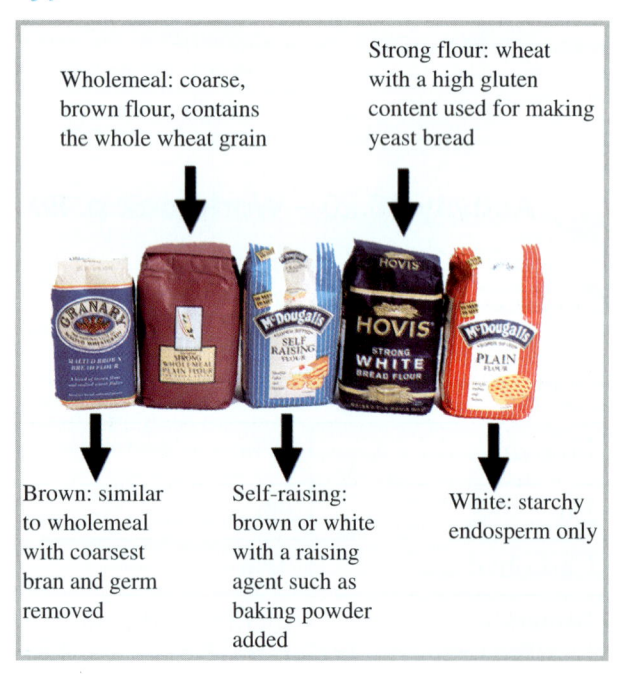

Wholemeal: coarse, brown flour, contains the whole wheat grain

Strong flour: wheat with a high gluten content used for making yeast bread

Brown: similar to wholemeal with coarsest bran and germ removed

Self-raising: brown or white with a raising agent such as baking powder added

White: starchy endosperm only

THE FOOD GROUPS   10

Gluten-free symbol

Gluten-free symbol: when this symbol appears on a product it is suitable for coeliacs

Gluten-free: flour with gluten removed; used to make bread etc. for coeliacs

Bread

Breads from around the world

Bread is one of our staple foods. Nowadays breads from all over the world are available in our supermarkets. Some of these breads are pictured above.

Pasta

Special wheat called *durum* wheat is used to make pasta. Flour made from this wheat is mixed with water and egg to make a dough. Sometimes flavouring ingredients such as tomato or spinach are added at this stage. It is then shaped and dried. Fresh pasta is not totally dried and must be used within a few days. Dried pasta is fully dried and keeps for up to a year.

Types of pasta

1. Long fusilli
2, 8, 9, 12, 15, 17. Conchiglie shells
3, 10. Short fusilli
4, 16. Farfalle (bow tie pasta)
5. Penne
6, 11. Tagliatelle
7, 14. Spaghetti
13. Lasagne
18. Plain, tomato and spinach fusilli

✏️ **Activity 10.21 – Workbook p. 81**

Rice

Types
1. **Long grain rice**: served with savoury dishes, e.g. curry; cooks in twenty-five minutes
2. **Medium grain rice**: Italian rice used for dishes where rice is mixed through, e.g. risotto
3. **Short grain rice**: used in sweet puddings and desserts
4. **Whole grain/brown rice**: brown long grain rice, high in fibre as bran is not removed; takes forty minutes to cook
5. **Convenience rice products**: boil-in-the-bag, non-stick, cooked frozen rice

✏️ **Activity 10.22 – Workbook p. 81**

📝 **Exam time – Workbook p. 82 – Cereals**

FRUIT AND VEGETABLE GROUP

Eat four servings or more per day.

Fruit

Classification of fruit

Citrus:	oranges, lemons, grapefruit, limes, tangerines, satsumas
Berries:	strawberries, raspberries, blackberries, gooseberries, blackcurrants, redcurrants
Stone:	peaches, plums, apricots, nectarines, damsons, cherries
Hard:	pears, apples
Dried:	raisins, sultanas, currants, dates, prunes, apricots
Others:	bananas, rhubarb, melon, pineapple

Orange

Strawberry

Peach

Pear

Bag of raisins

Pineapple and banana

✏️ **Activity 10.23 – Workbook p. 84**

Average composition of fresh fruit

Higher level

Protein	trace
Fat	0%
Carbohydrate	5–20%
Minerals	Calcium
Vitamins	A and C
Water	80–90%

Activity 10.24 – Workbook p. 85

Food value

Fruit is very important in the diet for *three* main reasons:
1. Many fruits have a high vitamin C content
2. Many fruits are high in fibre
3. Most fruits are low in kilocalories

Also, although less importantly, some fruits contain a little *iron* and *calcium*. Highly coloured fruits such as peaches contain *carotene*, which is changed into vitamin A by the body. Fruit contains carbohydrate in the form of fruit sugar (fructose) and traces of protein and fat.

Activity 10.25 – Workbook p. 85

ESSENTIALS FOR LIVING

Uses of fruit in the diet

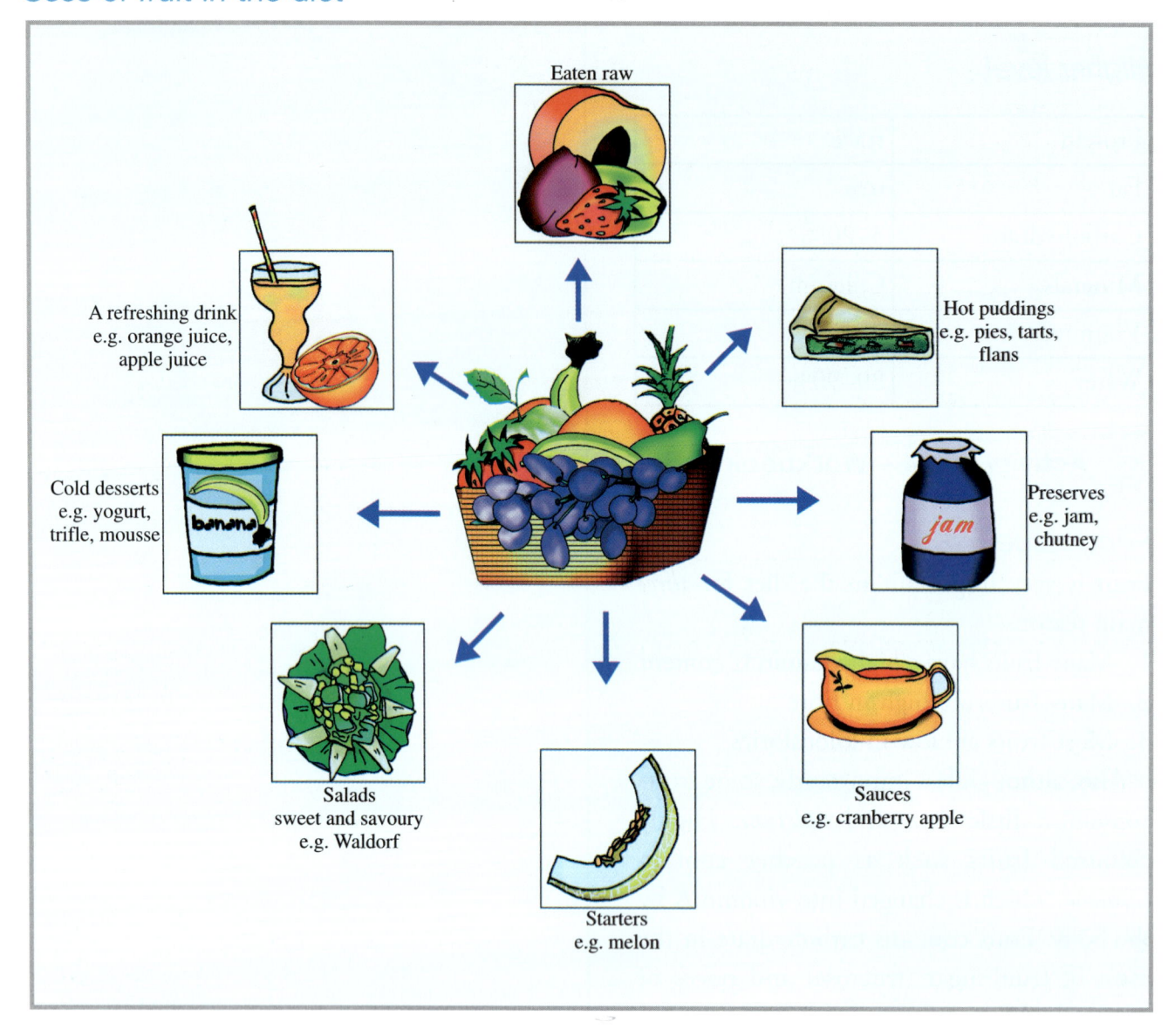

- Eaten raw
- Hot puddings e.g. pies, tarts, flans
- Preserves e.g. jam, chutney
- Sauces e.g. cranberry apple
- Starters e.g. melon
- Salads sweet and savoury e.g. Waldorf
- Cold desserts e.g. yogurt, trifle, mousse
- A refreshing drink e.g. orange juice, apple juice

Buying fruit	Storing fruit
• buy fresh undamaged fruit, heavy for size • buy medium-sized fruit: large lacks flavour • buy fruit in season • buy in small quantities, fruit spoils quickly • avoid pre-packaged fruit: you cannot examine quality	• remove plastic wrappings • store in vegetable drawer of fridge (not bananas) • use quickly • if fridge storage is not possible, store on a rack in a cool, dark, well ventilated place

Preparing fruit	Effects of cooking on fruit
• prepare just before cooking or eating to avoid loss of vitamin C • citrus: wash in warm water if rind is to be grated and eaten • hard fruits: cut in 4, remove core, leave skin on if possible; toss in orange juice to prevent browning, use this juice instead of sugar syrup in fruit salad • soft fruit e.g. strawberries: remove damaged fruit, wash gently in a sieve, pat dry with kitchen paper	• texture softens • vitamin C is reduced (up to 25%) • cellulose breaks down; fruit becomes more digestible; moulds and other micro-organisms are destroyed

Exam time – Workbook p. 85 – Fruit

Vegetables

Classification

Roots/tubers: carrot
parsnip
turnip (tuber)
beetroot
swede
onion (bulb)
potato (tuber)

Green: cabbage
lettuce
broccoli
spinach
Brussels sprout
kale
cauliflower

Pulses: peas
beans (green bean, broad bean, runner bean)
dried peas, beans, lentils

Fruits: tomato
pepper
cucumber
courgette
aubergine
pumpkin

Activity 10.26 – Workbook p. 86

Average composition of vegetables

Below are average figures for the composition of all vegetable types. For exact composition of specific ones, see the food tables in the workbook.

Protein	1–7% (pulses have most)
Fat	very little
Carbohydrate	2–20% (tubers have most)
Minerals	calcium, iron
Vitamins	vitamin A (carotene) in highly coloured vegetables; vitamin C, a little vitamin B
Water	75–95% (greens and fruits have highest)
Fibre	good source

Food value

Vegetables are valuable in the diet for *five* main reasons:

1. Pulse vegetables, that is, peas, beans and lentils, are a good source of vegetable protein. For this reason they are very important in the diet, especially of vegetarians.
2. Vegetables are low in fat and have a high water content. This means that they are low in kilocalories: good news for weight watchers and those at risk from heart disease.
3. Vegetables are a good source of vitamin C. Vitamin C is vital for general health and is a vitamin sometimes lacking in the Irish diet.
4. Vegetables, especially green vegetables, are a good source of iron. Iron is vital for healthy blood and is another nutrient often lacking in the Irish diet. In fact research shows up to seventy per cent of Irish teenage girls lack iron.
5. Vegetables are a great source of fibre. Fibre is needed to prevent constipation and other bowel problems.

Activity 10.27 – Workbook p. 87

Buying vegetables	Storing vegetables
• buy vegetables 'in season': they are the freshest and best • buy undamaged vegetables, heavy for size • buy medium-sized: large lack flavour • avoid pre-packed vegetables: damage may not be visible • buy in small quantities, they go off quickly	• use as soon as possible • remove plastic wrappings or bags • store greens and fruits e.g. peppers in the fridge • store roots and tubers in a cool, dark, ventilated place

In season: most vegetables grow at a certain time of year. This is when they are *in season* and at their best. Vegetables must be imported when they are *out of season*, and are usually expensive e.g. strawberries at Christmas.

Organically grown vegetables are vegetables grown naturally, without the use of artificial fertilisers. They are very expensive.

THE FOOD GROUPS

EU grading and regulations

Vegetables must be:
- sound and free from soil and chemicals
- graded according to size
- marked with the country of origin, variety and class

Classes:

Extra: excellent quality
Class I: good quality
Class II: marketable but with defects of shape or colour
Class III: marketable but inferior

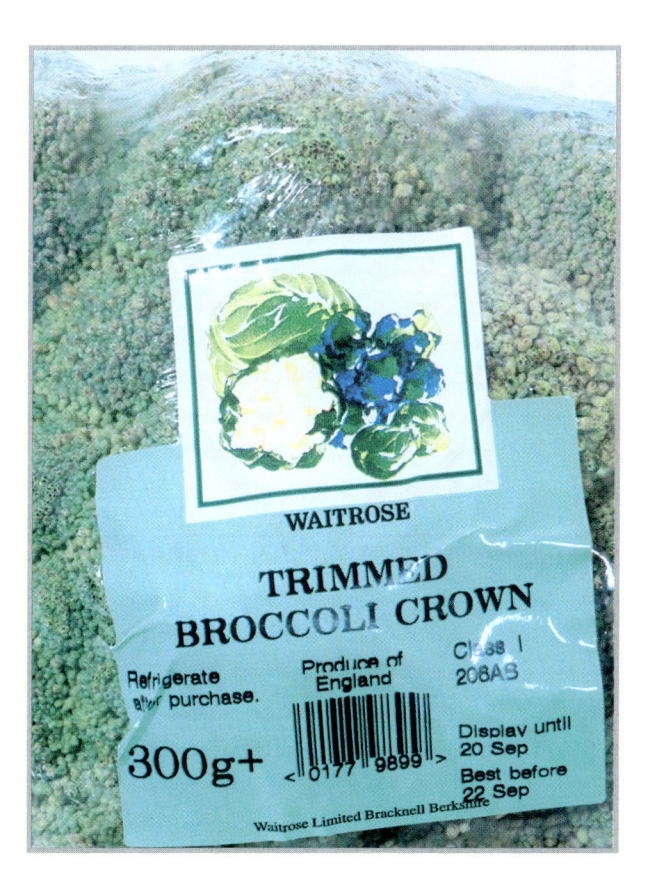

Cooking vegetables

Boiling: While not the best method of cooking vegetables, is traditional and still the most common way vegetables are cooked in this country.
- wash
- Remove
  - top and tail (roots)
  - withered leaves (greens)
  - pods (pulses)
- place in a *small* amount of boiling water

Cooking times (minutes)
- roots and tubers (10–20)
- greens (5–10)
- pulses (5–7)

Don't overcook

Vegetables may also be grilled (tomatoes), roasted (roots and tubers), steamed, stir-fried, stewed, pressure-cooked and microwaved.

Steaming: a healthy way to cook vegetables

ESSENTIALS FOR LIVING

Steaming: This is an excellent way of cooking vegetables. Because vegetables are not actually sitting in the cooking water, minerals and vitamins are not lost.

Stir-frying: Chop vegetables for stir-frying just before cooking. Cook in the minimum amount of oil (use oil spray). Vegetables are cooked quickly, nutrients are not lost.

Microwaving: Prepare vegetables and chop into even-sized pieces. Cover, cook on high for up to ten minutes (depends on type of vegetable). Allow vegetables to 'stand' for two to three minutes before serving.

Preventing nutrient loss when preparing and cooking vegetables

1. Buy good quality fruit and vegetables. Eat raw if possible.
2. Do not use bread soda: it destroys vitamin C. 
3. Use a small amount of water. Keep the lid on. Don't overcook (al dente: cooked but still has bite). 
4. Do not soak vegetables. 
5. Cook with the skins on if possible. If peeling, use a peeler: a knife peels too thickly. 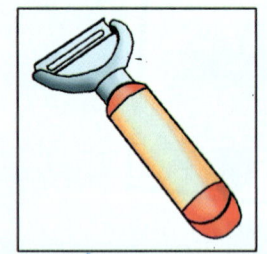
6. Chop with a sharp knife. 

Activity 10.28 – Workbook p. 87

Processed fruit and vegetables

Fruit and vegetables are generally processed to prolong their shelf life.

Frozen
- no loss of goodness but
- can go soft on thawing

Canned
- canning involves high temperatures: destroys vitamins C and B
- fruit canned in syrup has a high sugar content

Dried
- vitamins C and B are lost

Activity 10.29 – Workbook p. 87

Exam time – Workbook p. 89 – Vegetables

Revision crossword – Workbook p. 88

Now test yourself at *www.my-etest.com*

chapter 11

Breakfasts, packed meals, soups and sauces

BREAKFASTS

A healthy breakfast is an essential start to the day. When you wake up your blood sugar is low after fasting all night. If you don't eat a breakfast it remains low. Because of this you will be less able to concentrate at school or at work and more likely to eat high-calorie mid-morning snacks such as crisps and fizzy drinks.

Planning a healthy breakfast

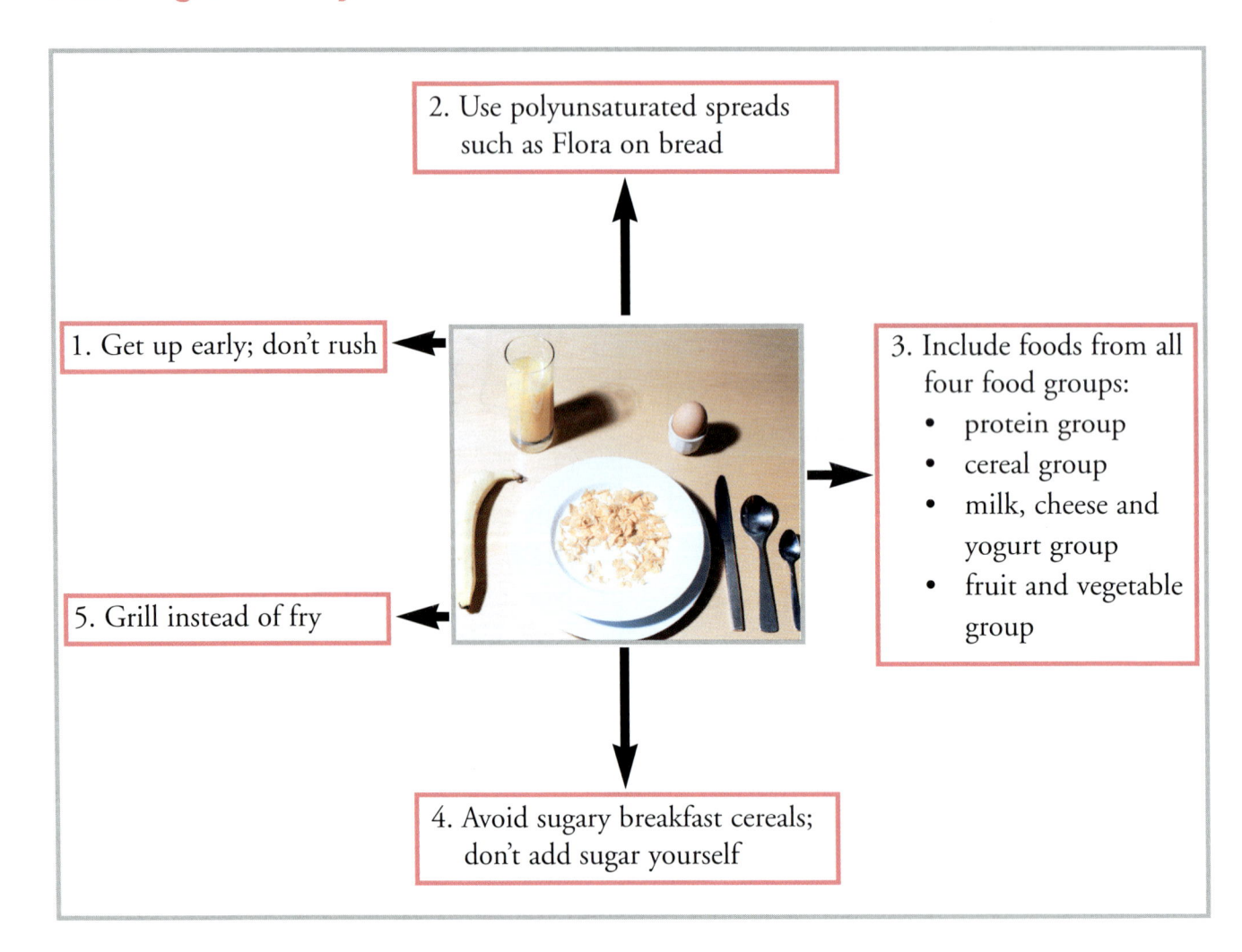

1. Get up early; don't rush
2. Use polyunsaturated spreads such as Flora on bread
3. Include foods from all four food groups:
   - protein group
   - cereal group
   - milk, cheese and yogurt group
   - fruit and vegetable group
4. Avoid sugary breakfast cereals; don't add sugar yourself
5. Grill instead of fry

BREAKFASTS, PACKED MEALS, SOUPS AND SAUCES   **11**

Choose foods from each group for a healthy breakfast.

1. *Fruit and vegetable group*
   - fruit juice (orange, grapefruit, cranberry)
   - grilled grapefruit
   - fruit segments
   - stewed fruit, e.g. prunes
   - grilled mushrooms or tomatoes

2. *Cereal and bread group*
   - healthy breakfast cereals include muesli, porridge with a sprinkle of All-Bran, cornflakes, bran flakes, Rice Krispies, Weetabix, Shredded Wheat, Fruit 'n Fibre
   - bread: brown, toast, scones, croissants, bagels

A healthy breakfast
Orange juice
Muesli with low-fat milk
Boiled egg and toast
Tea

3. *Protein group*
   - eggs (poached, scrambled, boiled, omelettes)
   - grilled bacon, sausage, pudding, kidney, liver
   - cold meats (ham, salami)

4. *Milk, cheese and yogurt group*
   - milk on cereal or alone
   - yogurt, cheese

ESSENTIALS FOR LIVING

Setting a breakfast tray

1. Collect all tableware: it should be spotless
2. Set tray as in diagram
3. Cover cooked main course with another plate to keep it warm

A breakfast tray

Key:
1. Toast rack/basket
2. Butter
3. Marmalade
4. Main course
5. Glass of fruit juice
6. Milk

Activity 11.1 – Workbook p. 92

Activity 11.2 – Workbook p. 93

Exam time – Workbook p. 94 – Breakfasts

PACKED MEALS, LUNCHES, PICNICS

Guidelines for a packed meal

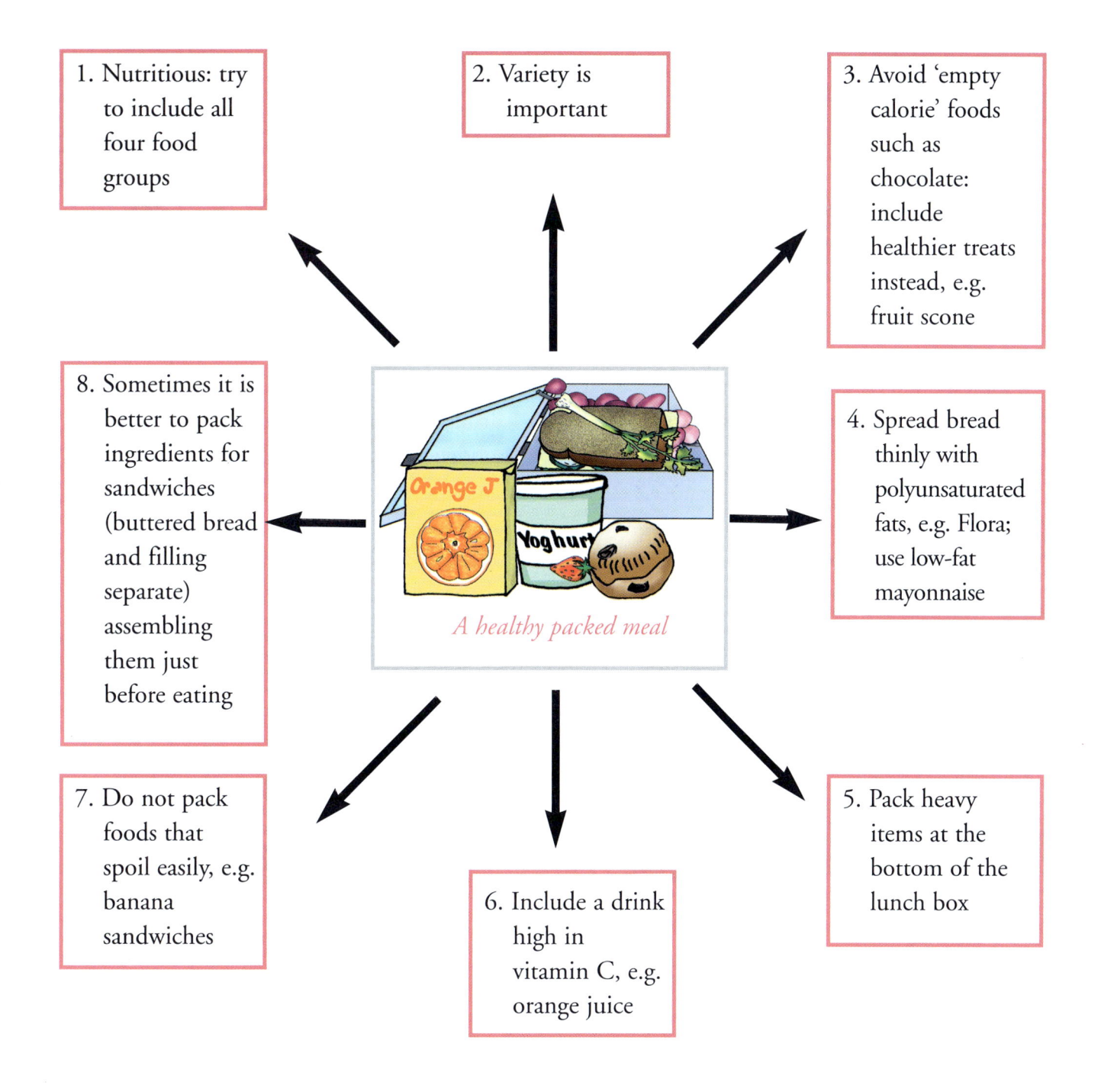

1. Nutritious: try to include all four food groups

2. Variety is important

3. Avoid 'empty calorie' foods such as chocolate: include healthier treats instead, e.g. fruit scone

4. Spread bread thinly with polyunsaturated fats, e.g. Flora; use low-fat mayonnaise

5. Pack heavy items at the bottom of the lunch box

6. Include a drink high in vitamin C, e.g. orange juice

7. Do not pack foods that spoil easily, e.g. banana sandwiches

8. Sometimes it is better to pack ingredients for sandwiches (buttered bread and filling separate) assembling them just before eating

A healthy packed meal

Designing a healthy packed meal

Choose foods from each group for a healthy packed meal.

Fruit and vegetable group
Apples, oranges, kiwi fruit, grapes, bananas, salads, fruit juice, soup, sticks of carrot, celery, Branston pickle

Milk, cheese and yogurt group
Milk, milk shakes, yogurt and yogurt drinks, cheese: alone, on crackers, or in sandwiches

Packed meal
Orange juice
Open wholemeal sandwich: ham, low-fat cheese, Branston pickle, Yogurt

Treats
Scones, light fruit cake or brack, digestive biscuits

Protein group
Sliced cold meats e.g. ham, chicken, beef, turkey; nuts, sausage rolls, chicken drumsticks, prawn cocktail, tuna, salmon

Cereal and potato group
Bread: rolls, baps, sliced wholemeal or white bread, croissants, bagels, rice and pasta salads

BREAKFASTS, PACKED MEALS, SOUPS AND SAUCES **11**

Sandwiches

Sandwiches need not be boring. They consist of three parts: bread, spread and filling, each of which can be varied. Make sure to use plenty of filling and that it is not too dry.

Sandwich ingredients

Activity 11.3 – Workbook p. 95

Exam time – Workbook p. 95 – Packed meals

Bread	Spread	Fillings
Wholemeal, soda, brown sliced, white sliced, rolls, baps, pitta pockets, bagels	Polyunsaturated margarine, low-fat spreads, spreadable butter, low-fat mayonnaise, salad cream, mustard, ketchup, Branston pickle, low-fat cream cheese, chutney	Ham, chicken, beef, turkey, luncheon meats, corned beef, fish: tuna, salmon, sardines, prawns; eggs, tomatoes, lettuce, scallions, onion, grated carrot, cucumber, peppers, bean sprouts, cress

Not only can the bread, spread and filling vary, but also the sandwich type.

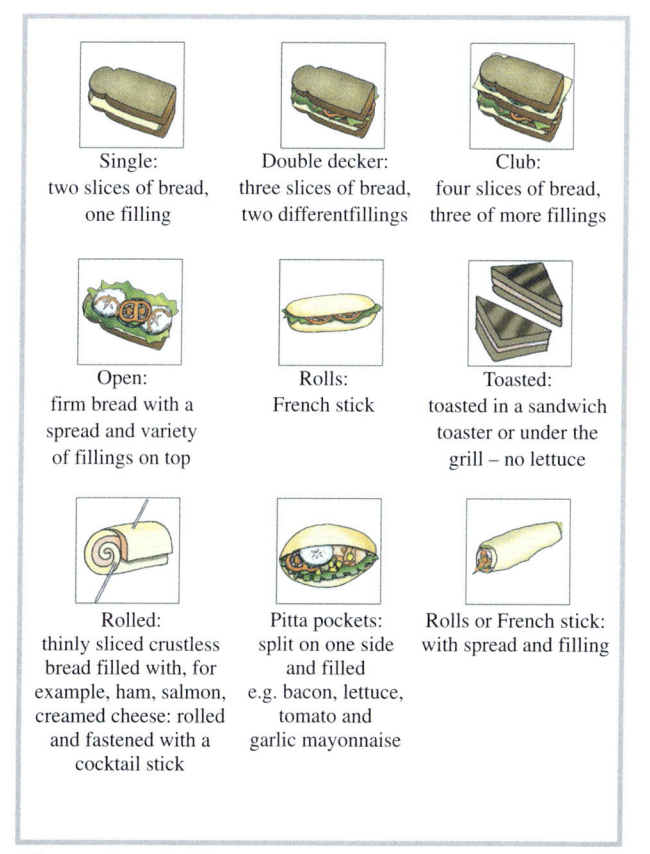

Single: two slices of bread, one filling

Double decker: three slices of bread, two different fillings

Club: four slices of bread, three of more fillings

Open: firm bread with a spread and variety of fillings on top

Rolls: French stick

Toasted: toasted in a sandwich toaster or under the grill – no lettuce

Rolled: thinly sliced crustless bread filled with, for example, ham, salmon, creamed cheese: rolled and fastened with a cocktail stick

Pitta pockets: split on one side and filled e.g. bacon, lettuce, tomato and garlic mayonnaise

Rolls or French stick: with spread and filling

Types of sandwiches

SOUPS

Soups can be classified into thick soups and thin soups.

Thick soups

Puréed: thickened by blending the soup's own ingredients using a sieve, liquidiser or food processor e.g. tomato soup

Thickened: by adding a starch, e.g. flour, cornflour or pasta

Thin soups

Clear soup: thin soup made with well flavoured stock

Broth: thin clear soup with finely chopped meat, vegetables and a starch such as barley floating in it. For example, chicken broth is made from a thin clear stock, with chicken pieces, vegetables and perhaps barley.

A well garnished soup

A good soup . . .

- is made from a fresh, well flavoured stock: soup should be seasoned with pepper and herbs; do not add too much salt – use a salt substitute
- tastes of its main ingredient, e.g. tomato soup should have a strong tomato flavour
- is piping hot and has no grease floating on the top (*Note:* some soups are meant to be served cold)
- has a good colour

A thick soup should not be too thick and should not have starchy lumps.

Three ways to thicken a soup

1. Blend 25 g of flour or cornflour with cold water. Stir this mixture into the soup just before the end of cooking. Bring soup back to the boil and boil for approximately five minutes. Garnish and serve.
2. Begin by gently frying (sautéing) the soup ingredients, e.g. vegetables, in 25 g of fat or oil. Add 25 g flour, cook for a few minutes. Gradually add the stock. Soup will thicken once it comes to the boil.
3. Add barley, rice or pasta to soup twenty minutes before the end of cooking.

BREAKFASTS, PACKED MEALS, SOUPS AND SAUCES 11

How to garnish a soup

Croutons (cubes of bread fried in oil – use polyunsaturated)

Herbs, e.g. parsley

Swirl of cream, e.g. cream of vegetable soup

Convenience soups

Dried packet soup

- some are instant, e.g. cup-a-soup; while cheap they are not very nutritious

Canned soup

- just needs heating up; very convenient

ESSENTIALS FOR LIVING

Fresh cartons

- expensive but tasty and nutritious; store in the fridge

Activity 11.4 – Workbook p. 96

SAUCES

Sauces are well flavoured liquids and can take many forms. They can be sweet or savoury, hot or cold. Sauces can be used as a garnish or can be part of the dish itself, for instance a brown sauce is part of a brown stew. Sauces can also bind or coat foods and generally add colour and flavour to a dish.

Classification

Type	Examples
1. roux	white, cheese
2. fruit purée	apple, raspberry coulis, cranberry
3. egg	egg custard
4. cold	mint
5. other	chocolate, butterscotch, barbecue

Roux

A roux sauce is made from equal amounts of fat and flour. Different flavourings and amounts of liquid are added to vary the basic sauce.

There are four basic thicknesses of roux sauces. The amounts of fat and flour stay the same for each one. It is the amount of liquid that changes.

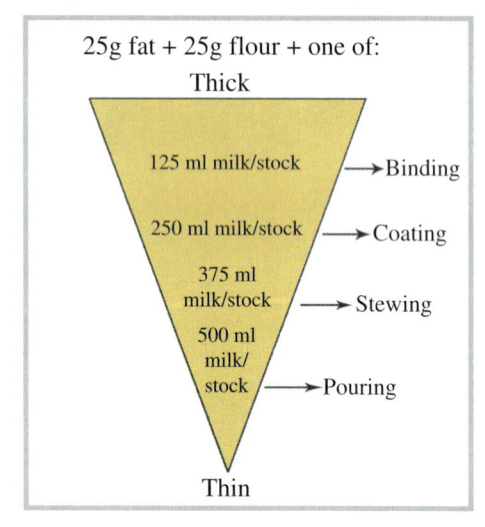

25g fat + 25g flour + one of:
Thick
- 125 ml milk/stock → Binding
- 250 ml milk/stock → Coating
- 375 ml milk/stock → Stewing
- 500 ml milk/stock → Pouring
Thin

Note: if a recipe says, for example, '500 ml of coating sauce' it is the amount of liquid that is meant. So ingredients for 500 ml of coating sauce are 50 g fat, 50 g flour, 500 ml liquid.

106

BREAKFASTS, PACKED MEALS, SOUPS AND SAUCES | 11

How to make a basic white roux sauce

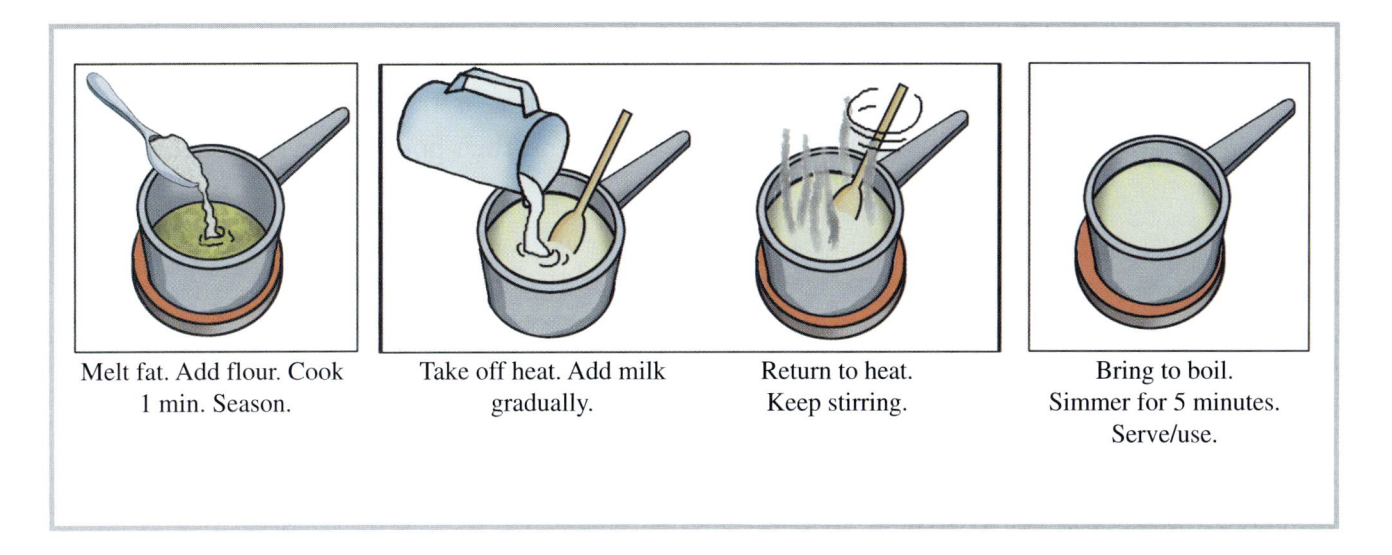

| Melt fat. Add flour. Cook 1 min. Season. | Take off heat. Add milk gradually. | Return to heat. Keep stirring. | Bring to boil. Simmer for 5 minutes. Serve/use. |

Variations on basic pouring sauce (500 ml)

Sauce	Extra ingredients	Goes well with
Parsley	add 2 teaspoons chopped parsley	bacon, fish
Mustard	add 1 teaspoon made mustard, 2 teaspoons vinegar	fish
Cheese	¼ teaspoon mustard (added to roux), 50 g grated cheese added at the end and allowed to melt: do not return to heat	fish, cauliflower
Mushroom	Add 50 g sautéed mushrooms after liquid has been added and then simmer for 5 minutes	steak, roast beef
Pepper	Fry 1 teaspoon crushed black pepper in margarine before adding flour and milk	steak, chicken

107

Fruit coulis

Desserts look very well if you decorate them with a simple fruit coulis. To make the coulis simply purée highly coloured fruits such as strawberries or raspberries, add a little icing sugar and then pour to the side of the dish.

✏️ **Activity 11.5 – Workbook p. 97**

✏️ **Activity 11.6 – Workbook p. 98**

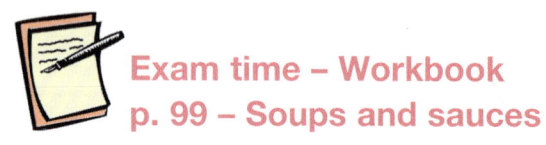

📝 **Exam time – Workbook p. 99 – Soups and sauces**

Now test yourself at *www.my-etest.com*

chapter 12

Food processing, food preservation and convenience foods

Food processing

Food processing means treating foods in some way to make them easier to use. Most foods nowadays undergo some degree of processing.

Before processing	After processing
Wheat grains	flour
Unpasteurised milk	pasteurised milk
Milk	cheese
Raw chicken	chicken curry, TV dinner
Pork	frozen sausage rolls
Fresh vegetables	packet of frozen vegetables
Mandarin oranges	tin of mandarin oranges
Potatoes	oven chips

Food preservation

Food preservation is a form of food processing that slows down food spoilage (food going bad). Food spoilage is caused by either *enzymes, moulds, yeast* or *bacteria* attacking food. Spoilage of some foods, such as butter, may be caused by oxygen in the air.

Enzymes, moulds, yeast and bacteria need *five conditions* to grow:

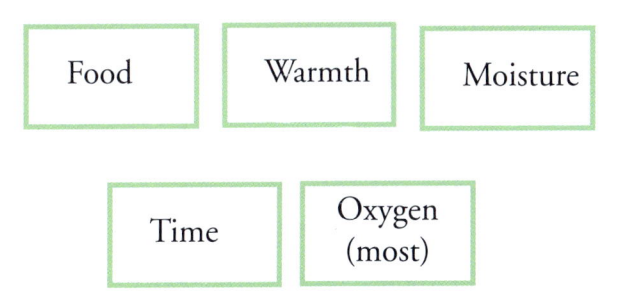

Food   Warmth   Moisture

Time   Oxygen (most)

Take one or more of these conditions away *or* add a chemical preservative, such as vinegar, and the food is preserved.

ESSENTIALS FOR LIVING

METHODS OF PRESERVATION

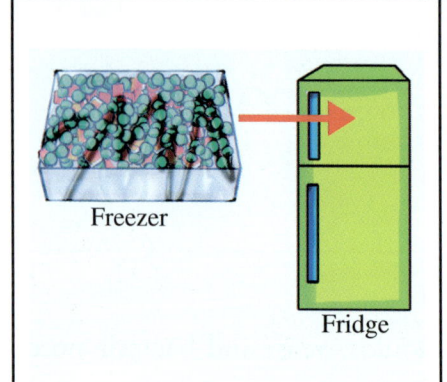 freezing, refrigeration = removing warmth	 pasteurisation, canning, bottling = applying strong heat	 drying = removing moisture
 canning / bottling = removal of air	 adding chemical preservatives e.g. sugar, vinegar, smoke, salt, other 'E' numbers	irradiation – passing rays through foods (usually fruit and vegetables); rays destroy enzymes and bacteria so food keeps longer; it is not known if this method is safe

Freezing

Food is brought to a temperature of −18°C or below. Enzymes and micro-organisms are not killed at this temperature but become inactive (sleep). Once food is *thawed* enzymes and micro-organisms become active again.

Guidelines for successful freezing

Frozen food, if frozen properly, is almost as nutritious as fresh food. Because food is usually frozen while at its freshest and most nutritious it is in fact more nutritious than fresh food that has been lying around for a few days.

The first rule of freezing is turn the freezer to its coldest setting two to three hours before freezing and freeze food *quickly*.

110

FOOD PROCESSING, FOOD PRESERVATION AND CONVENIENCE FOODS | 12

- freeze food at its freshest
- don't freeze too much at a time
- cool food before freezing it
- choose suitable packaging, e.g. plastic freezer bags, plastic tubs
- freeze in usable amounts
- if there is a 'fast freeze' cabinet use it to freeze foods
- label foods clearly

Quick freezing: small ice crystals are formed: on thawing, cell walls of food are unbroken, food stays firm and keeps its shape

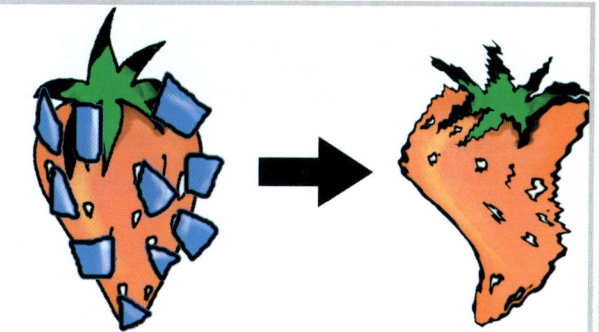

Slow freezing: large ice crystals are formed: on thawing, food loses its shape and nutrients and becomes soft and mushy

Buying frozen foods	Thawing frozen foods
• packaging untorn • frozen solid • shop's freezer should read −18°C or less • check food is below the 'load line' in open freezers	• some foods can be cooked from frozen e.g. burgers, chips: read the packet instructions • thaw meat completely, especially chicken • food can be thawed in the microwave or in the fridge overnight • *never* refreeze thawed food

Blanching

Blanching is done to destroy enzymes that can reduce the quality of frozen vegetables.

Plunge prepared vegetables into boiling water for 1–4 minutes

Plunge into ice-cold water

Blanching times

1 minute	mushrooms, peas
3 minutes	broccoli, parsnips, turnips, cauliflower, beans
4 minutes	carrots, Brussels sprouts

111

ESSENTIALS FOR LIVING

Open freezing

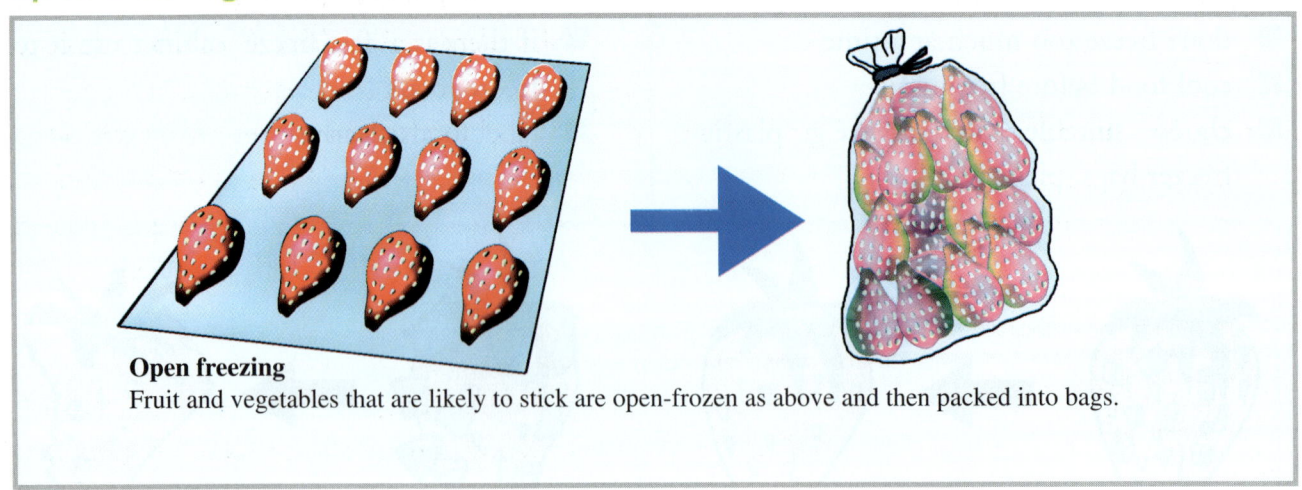

Open freezing
Fruit and vegetables that are likely to stick are open-frozen as above and then packed into bags.

Blast freezing

This is a commercial method of freezing. Very cold air (–35ºC) is blown over food to freeze it quickly.

✎ **Activity 12.1 – Workbook p. 100**

📝 **Exam time – Workbook p. 100 – Preservation**

Other methods of preservation

Drying	Chemical preservation	Canning and bottling
• enzymes and micro-organisms need moisture (water) to survive • dehydration or drying removes moisture and so preserves the food • examples: breakfast cereals, pasta, rice, raisins (dried grapes), soups	• certain chemicals preserve food; many of them are natural and have been used for centuries • sugar: jam • vinegar: pickles • salt: canned fish • smoke: fish • there are also many artificial preservatives	• food is heated to a high temperature (this kills enzymes and micro-organisms) • food is then sealed in sterile containers • food keeps for years • vitamins B and C are lost • don't buy damaged or bulging cans (they may contain food poisoning bacteria)

✎ **Activity 12.2 – Workbook p. 101** ✎ **Activity 12.3 – Workbook p. 102**

FOOD PROCESSING, FOOD PRESERVATION AND CONVENIENCE FOODS

FOOD LABELLING

Higher level

Under EU law all packaged food must have certain information on the label. These items of information are illustrated below.

- Name of food
- List of ingredients: product's main ingredient (by weight) listed first, etc.
- Net quantity
- Use by/best before/sell by date
- Storage instructions
- Name and address of manufacturer
- Country of origin
- Instructions for use (if necessary)
- Nutritional information (if product is making health claims)

FOOD ADDITIVES

Higher level

In 1984 the EU decided to make it law for food manufacturers to display lists of ingredients (including all additives) on food packaging. A system of coding additives was also developed and many additives were given an *E number*.

There has been much public debate and misunderstanding about E numbers since this time. Many people believe all E numbers to be a health risk, causing hyperactivity and allergic reactions. It is because of this bad press that most food labels now list additives by their full names instead of by their E numbers.

What are the facts?

Additives *thought* to be safe by the EU's scientific committee for food are given an E number. Many E numbers are natural healthy substances, for example vitamin C is E300, vinegar is E260. There are, however, questions about the safety of some additives still being passed as safe under EU regulations (see tartrazine and monosodium glutamate below). Another important issue is that of the *cocktail effect*. This means that while individual additives may be quite safe on their own they may not be so when mixed with others.

Tartrazine is a legal yellow food colouring. It is added to some custards, ice creams, jams, cereals and soft drinks. Some studies strongly link tartrazine to hyperactivity and allergies in children.

Monosodium Glutamate (MSG) is another controversial additive. MSG is a white, crystalline substance used in the preparation and processing of many foods. Although MSG has a mild flavour itself, it functions by bringing out the tastes of such foods as meats, vegetables, seafood, soups, sauces, and casseroles. It is called a *flavour enhancer* and works by stimulating the taste buds, thus strengthening the taste of the food. MSG has been used for many centuries in oriental and Asian cooking (e.g. Chinese and Indian).

Although many studies have found MSG to be perfectly safe, like tartrazine it has been linked to hyperactivity and allergies in children. Another criticism of MSG is that once people get used to highly flavoured MSG foods they will dislike milder, often healthier, alternatives.

Food additives and their functions

Type of additive	What it does
Colourings (E100 – E199)	make food look more appetising
Preservatives (E200 – E299)	prevent food from going off, e.g. vinegar preserves pickled onions
Antioxidants (E300 – E399)	these prevent fats, e.g. butter, reacting with the air and going rancid (off)
Emulsifiers and stabilisers (E400 – E499)	emulsifiers allow oil and water to mix; stabilisers keep mixtures from separating, e.g. mayonnaise
Flavourings (no E number)	add to the flavour of food e.g. vanilla essence
Flavour enhancers (E600+)	strengthen the flavour of food e.g. MSG
Sweeteners	e.g. saccharin
Nutritional additives (no E number)	these improve the nutritional quality of food or replace nutrients lost in processing

CONVENIENCE FOODS

Convenience foods are foods that have been processed in some way so as to make them easier and quicker to use. There are basically two types of convenience foods:
1. Single product foods, e.g. can of tomatoes, frozen vegetables
2. Complete meals (often called cook-chill foods), e.g. lasagne

Single product convenience foods

These products can be combined very successfully with fresh foods to make very nourishing meals. For example grilled lamb chops with freshly boiled potatoes and frozen vegetables.

Activity 12.4 – Workbook p. 103

Complete meals

Over the past number of years a huge number of new 'complete meal' products have come on the market. When these products first came on the market in the 1980s their quality was not always good. With advances in technology both in the food industry and in the home these products are now of a much higher standard.

While some of these products are dried, for example pot noodles, many of the better ones are what are called *cook-chill foods*. Basically these complete meals are prepared, cooked and frozen or chilled in the factory. All the consumer has to do is take them home, heat them up and eat!

The dangers of cook-chill foods
Cook-chill foods, because they are only reheated and not cooked again at home, can be a source of food poisoning. These foods are safe if they are stored in a fridge or freezer below 4°C and are reheated thoroughly. If a microwave is used to reheat the food it must be stirred so that all parts of the food reach a high enough temperature. Cook-chill foods are expensive and sometimes portions are small.

Activity 12.5 – Workbook p. 103

Activity 12.6 – Workbook p. 104

GENETICALLY MODIFIED FOODS (GM FOODS)

Genetically modified foods or GM foods, as they are called, are foods that have been scientifically changed to resist disease, yield bigger, less flawed crops etc. Many environmental groups, for example Greenpeace, have been actively campaigning against GM foods for many years. These groups worry about the long-term effects of this 'tampering with nature' on us and our environment. Sometimes activists like the one pictured here resort to destroying GM crops by way of protest.

Exam time – Workbook p. 105 – Convenience foods

Revision crossword – Workbook p. 105

Now test yourself at *www.my-etest.com*

chapter 13

Home baking

The main advantage of baking your own bread, cakes and biscuits is that you can control exactly what goes into them. Many commercially produced cakes and buns contain a high proportion of fat and sugar. It is possible to bake at home using less fat or unsaturated fat and less sugar too.

GUIDELINES FOR HOME BAKING

- collect all ingredients
- weigh accurately

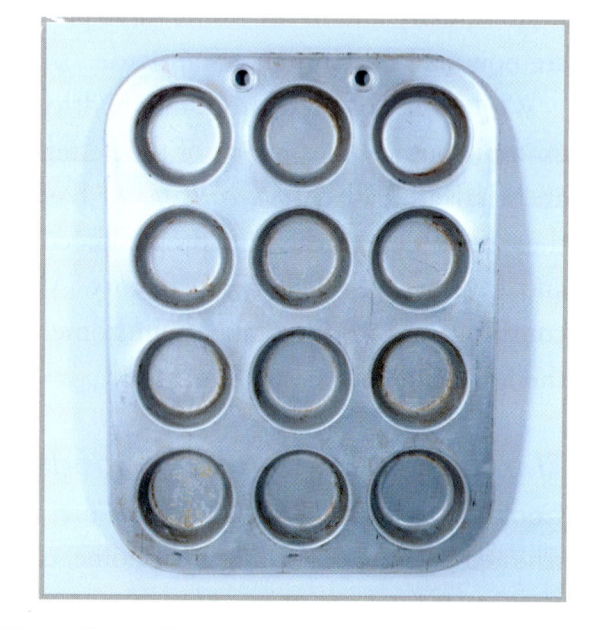

- collect all equipment
- prepare tins

HOME BAKING 13

- arrange oven shelves, *then*
- set oven temperature accurately
- preheat fully (electric ovens: light goes off)
- time cooking accurately
- don't open oven door unnecessarily

RAISING AGENTS

A raising agent is something that makes bread and cakes rise.

There are four raising agents:

1. Air
2. Bread soda
3. Baking powder
4. Yeast

Baking powder

Yeast

How they work

Air

Air is introduced into the mixture by sieving, rubbing in, creaming or whisking.

117

ESSENTIALS FOR LIVING

Once the mixture is heated the hot air in the mixture rises, pushing the mixture up. A crust forms on top, which stops the mixture from collapsing when cool.

Baking powder and bread soda

These two raising agents cause a chemical reaction in the mixture.

before it goes into the oven; a crust then forms in the oven keeping it in its risen state.

✏️ Activity 13.1 – Workbook p. 107

✏️ Activity 13.2 – Workbook p. 107

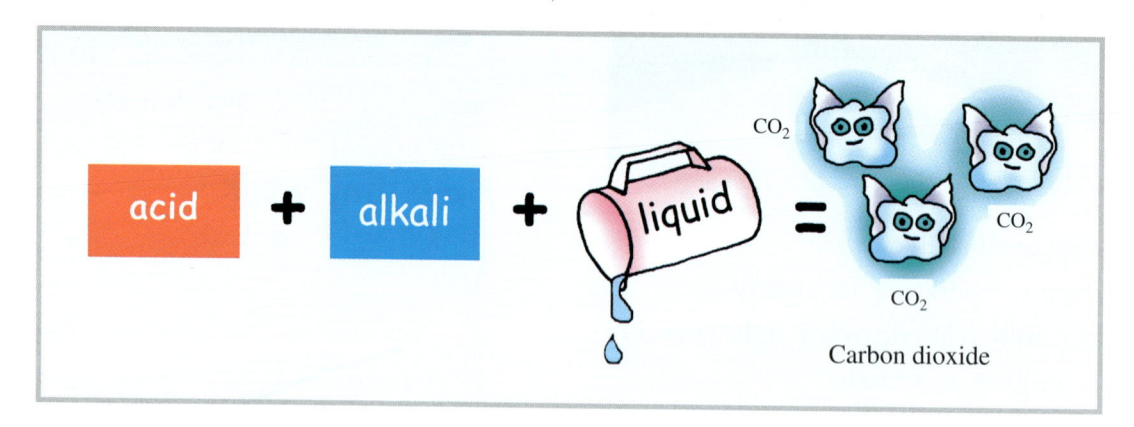

acid + alkali + liquid = CO_2 Carbon dioxide

Bread soda is an *alkali*; when it is mixed with an acid liquid such as buttermilk it produces CO_2.

Baking powder contains both an acid and an alkali. When liquid is added, for example when you add egg, CO_2 is produced.

The CO_2 produced works much like the air described above in that it pushes the mixture up. In the oven a crust forms, which keeps the mixture risen.

Note: Once the acid and the alkali get wet they begin producing CO_2. It is important to get the mixture into the oven quickly or the CO_2 will escape before the crust forms and the mixture will not rise properly.

Yeast

Yeast are living organisms used in bread-making. When they are warm and moist (in bread dough) they produce CO_2. Yeast dough rises

METHODS OF MAKING BREAD AND CAKES

Rubbing in

Fat is rubbed into the flour with the tips of the fingers until it looks like breadcrumbs. Liquid is then added. Example: scones.

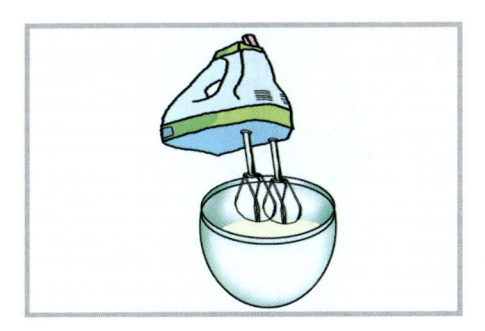

Creaming

Fat and sugar are beaten together until white and creamy. Egg, flour and raising agent are then added. Example: Queen cakes.

Whisking

Eggs and sugar are whisked until thick and creamy. Flour is then gently folded in. Example: sponge cake.

Melting

All ingredients that melt are melted together, e.g. fat, sugar, syrup. This mixture is then added to the flour and other dry ingredients. Example: gingerbread.

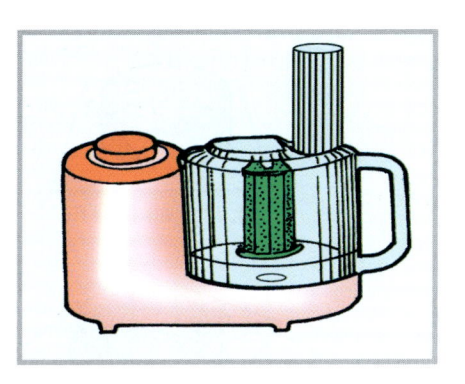

All-in-one

All ingredients are added at once. They are often mixed in a food processor. Example: all-in-one Madeira mix.

Activity 13.3 – Workbook p. 107

Tin preparation

Food	Preparing the tin
bread, plain scones (no sugar)	dust lightly with flour
pastry, small cakes and buns e.g. coconut buns	grease lightly
fatless sponge	grease lightly and dust with a mixture of castor sugar and flour
light cakes e.g. Madeira	line base only, grease sides
richer cakes	line with greaseproof paper (see diagrams below); dabbing some oil on the tin itself will make the greaseproof paper fit more snugly

ESSENTIALS FOR LIVING

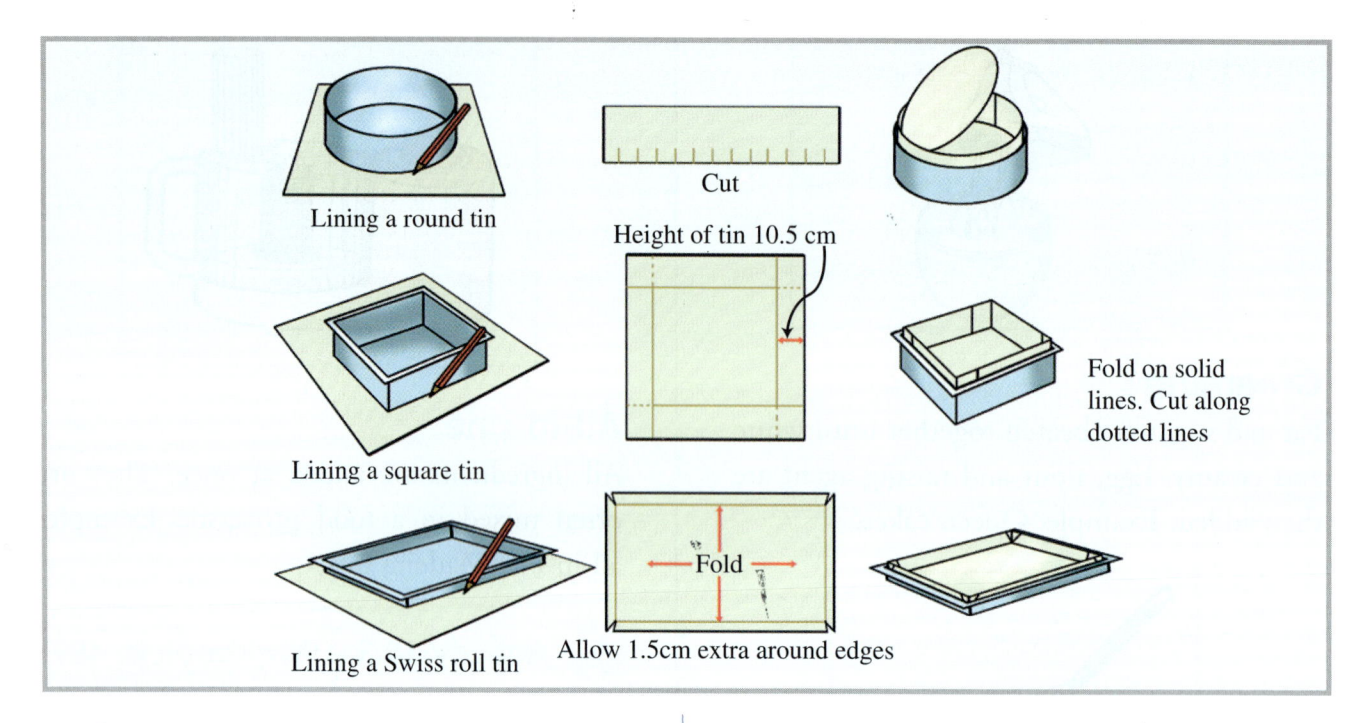

Lining a round tin

Cut
Height of tin 10.5 cm

Fold on solid lines. Cut along dotted lines

Lining a square tin

Lining a Swiss roll tin

Fold
Allow 1.5cm extra around edges

PASTRY

Types of pastry and their uses

Type	Uses	Description
short-crust and wholemeal short-crust	tarts, pies, quiche, sausage rolls	half the amount of fat to flour; just add water. Add wholemeal flour to make wholemeal pastry; increases fibre content
rich short-crust	mince pies, sweet flans, pies e.g. lemon meringue pie	short-crust pastry made rich by adding icing sugar and eggs
cheese pastry	quiche	add grated cheese to short-crust pastry
choux pastry	éclairs and profiteroles	melt fat in water, add flour and cook for a few minutes, then gradually beat in egg; spoon or pipe onto tin
flaky, puff and rough puff	vol-au-vents, sausage rolls	special rolling and folding techniques are used to introduce air into the pastry; high in fat, very time consuming
filo	spring rolls	very thin Greek pastry

HOME BAKING

Pastry-making guidelines

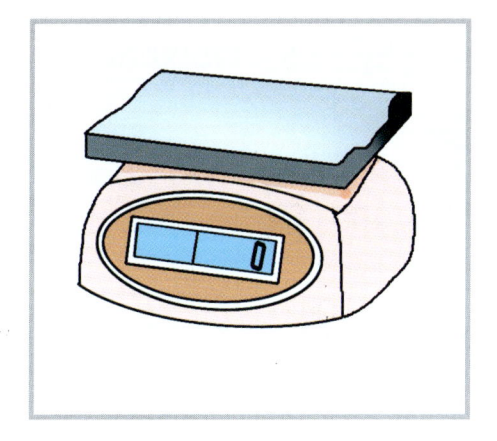

- weigh accurately
- keep everything cold 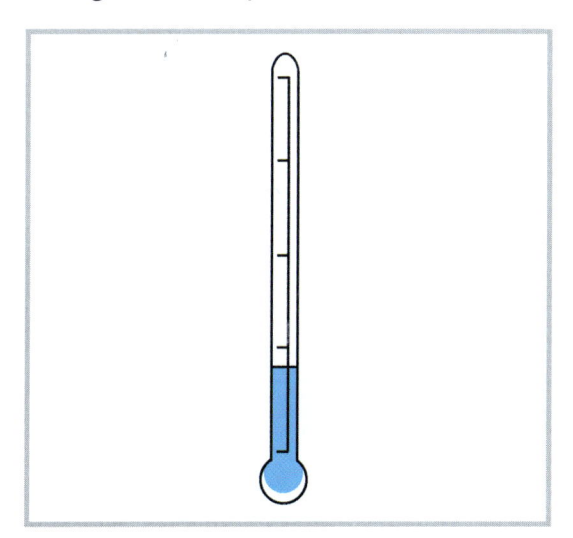
- introduce air (e.g. by sieving, rubbing in, rolling and folding) 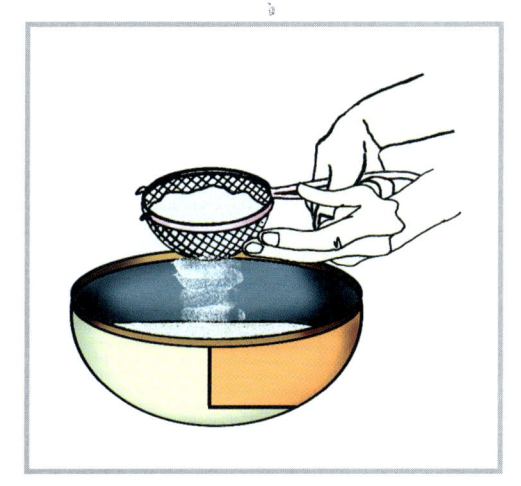
- handle pastry as little as possible 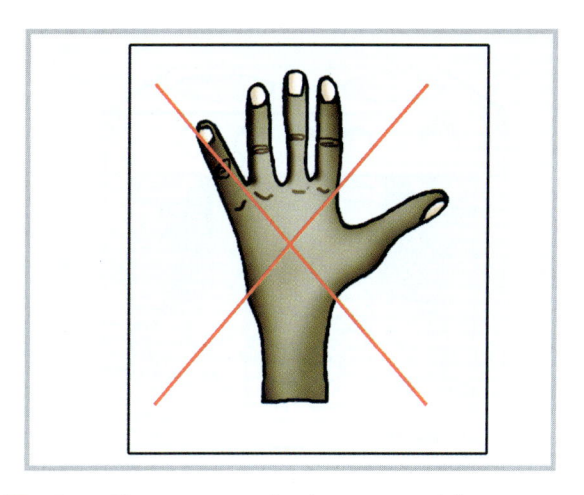
- roll lightly in one direction only on a lightly floured board; turn pastry to prevent sticking; do not stretch pastry 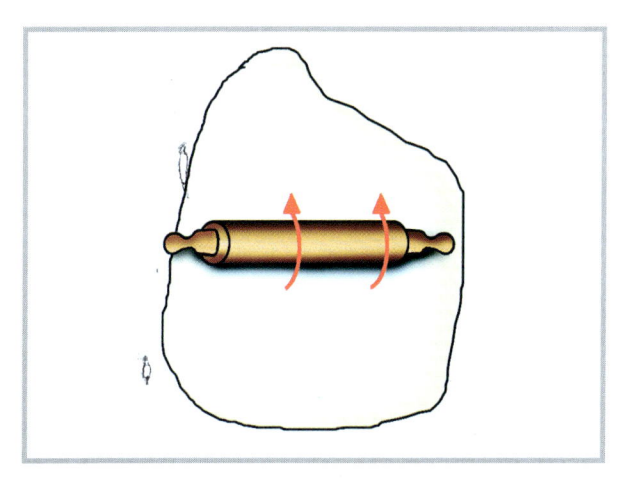
- allow pastry to relax in fridge before rolling

ESSENTIALS FOR LIVING

- bake first in a hot preheated oven (220°C/ Gas 7) then reduce the heat to finish baking

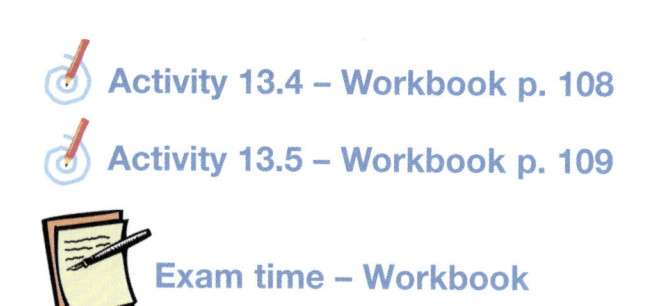

Activity 13.4 – Workbook p. 108

Activity 13.5 – Workbook p. 109

Exam time – Workbook p. 109

Now test yourself at *www.my-etest.com*

chapter 14

Recipes

	Page
RECIPE INDEX	

Breakfasts
Grilled grapefruit	124
Scrambled eggs	125
French savoury omelette	125

Starters
Chilled melon and summer fruits	126
Apple and walnut stuffed avocado	126
Garlic bread	127

Soups
Farmhouse vegetable	127
Cream of mushroom	128
Tomato and basil	128

Snacks
Croque monsieur	129
French toast sandwiches	129
French bread pizza	130

	Page
MAIN COURSES	

Beef, pork and lamb
Spaghetti bolognese	130
Lasagne	131
Beef or lamb curry	132
Home-made burgers	132
Stuffed pork chops	133
Stuffed liver and bacon	133

Chicken
Fruity chicken curry	134
Chicken and bacon risotto	134
Sweet and sour chicken or pork	135

Fish
Tomato and fish bake	136
Fish cakes	136

Milk, eggs and cheese
Quiche Lorraine	137
Pizza	138

Vegetarian
Vegetable stir-fry with rice	138
Vegetable curry	139
Cheesy vegetable and pasta bake	140

ESSENTIALS FOR LIVING

	Page
DESSERTS	
Fresh fruit salad	141
Apple and plum crumble	141
Bread and butter pudding	142
Delicious filled pancakes	142
Rich chocolate orange sauce	143
Butterscotch sauce	143
Fruit coulis	144

PASTRY

Short-crust pastry (also wholemeal and cheese)	144
Sausage rolls	145
Mince pies	145

BREAD AND CAKES

Rubbing-in method

Brown bread	146
Tea scones (also brown and fruit)	146
Banana bread	147
Coconut buns	147
Rock buns	148

Creaming method

Queen cakes	148
Welsh cheese cakes	149

All-in-one method

All-in-one chocolate cake	149
Simple carrot cake	150

	Page
Whisking method	
Fresh strawberry sponge cake	151
Melting method	
Ginger snaps	151
Gingerbread	152

BREAKFASTS

Grilled grapefruit

(serves 2)

Ingredients

- 1 grapefruit
- 1 teaspoon brown sugar
- 1 cherry

Grapefruit cut around equator

Method

1. Wash and cut the grapefruit around the equator.
2. Use a grapefruit knife to separate flesh from pith. Cut into segments.
3. Sprinkle with a teaspoon of brown sugar. Place under grill until sugar caramelises. Decorate with cherry.

Work card 1: Workbook p. 111

124

Scrambled eggs

(serves 2)

Ingredients	
• 3 large eggs • 3 tablespoons milk • 15 g butter or margarine • salt and pepper	*Garnish:* parsley *Serve with:* toast

Method
1. Make toast.
2. Beat eggs, salt and pepper in a bowl.
3. Heat milk and butter in a saucepan until butter melts. Add eggs.
4. Stir over a low heat until it becomes thick and creamy.
5. Do not overcook. Serve on warm toast, garnish with parsley.

Top Tip:
Add grated cheese, cooked rasher or ham for more texture and taste.

Work card 2: Workbook p. 111

French savoury omelette

(serves 1)

Ingredients	
• 2 large eggs • 15 g butter or margarine • 1 tablespoon water • salt and pepper • mixed herbs • oil to fry	*Fillings:* • choose 1–2 • 25 g grated cheese • 1 grilled rasher, chopped or • 1 slice ham, chopped • 50 g sautéed mushrooms

Method
1. Beat eggs, butter, water and seasonings in a bowl.
2. Heat oil in a small pan. Pour egg mixture in. As egg sets pull it gently towards the middle of the pan with a palette knife. Tilt the pan and run the uncooked egg into it. Continue until egg is just set.
3. Add your filling of choice.
4. Using the palette knife, fold the omelette over.
5. Slide onto a warm plate. Serve with toast, garnish with parsley.

Work card 3: Workbook p. 112

STARTERS

Chilled melon and summer fruits
(serves 2)

Ingredients	
• 1 small melon • 4 strawberries • 8 raspberries	• castor sugar to decorate wine glass

Method

1. Cut the melon along the equator (large melons are cut from top to bottom). Remove the seeds.
2. Using melon baller, scoop out the flesh.
3. Wash and slice strawberries, leave the raspberries whole.
4. Place the fruit in a wine glass, chill and then serve glass on a small plate with a doily.

Top Tip:
Decorate the wine glass first by dipping the rim in water and then in castor sugar.

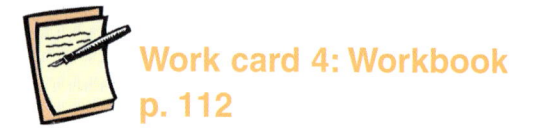

Work card 4: Workbook p. 112

Apple and walnut stuffed avocado
(serves 4)

Ingredients	
• 2 ripe avocado pears • 1 red eating apple • 50 g walnuts • 2 tablespoons low-fat mayonnaise	• lemon juice *Garnish:* walnut halves *Serve with:* lettuce, cherry tomatoes, cucumber slices

Method

1. Prepare the avocados:
   - Wash and cut into avocados all round until you reach the stone.
   - Take half the avocado in each hand and twist to separate.
   - Remove stone by piercing with a knife and twisting.
   - Remove flesh from skin in one piece with a dessertspoon.
   - Chop into even-sized pieces. Sprinkle with lemon juice to prevent browning.
2. Wash, core and chop apple. Chop walnuts.
3. Mix avocado, apple and walnuts with low-fat mayonnaise.
4. Fill avocado shells with mixture.
5. Garnish with walnut halves. Serve with lettuce leaves, cherry tomatoes and cucumber slices.

Top Tip:
To make cucumber look nice: before you slice it, cut into the cucumber by dragging a fork along its length. When you slice it you get a nice edge.

RECIPES 14

Cucumber with serrated edge

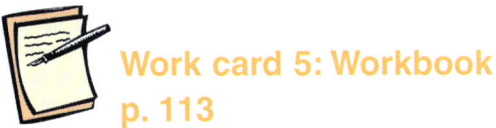

Work card 5: Workbook p. 113

Garlic bread

Ingredients	
• 1 French stick • 2 cloves garlic • 75 g butter or low-fat spread • 1 teaspoon chopped parsley	*Garnish:* sprig of parsley

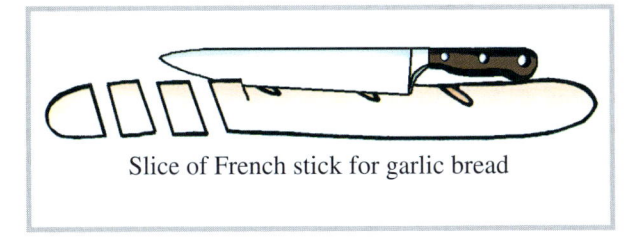

Slice of French stick for garlic bread

Method
1. Peel and crush garlic. Mix with butter and some parsley.
2. Slice bread as shown. Butter each slice.

Put stick back together and wrap in foil.
3. Bake at 200°C / Fan 190°C / Gas 6 for 15 minutes.
4. Remove foil. Serve hot, garnished with a sprig of parsley.

SOUPS

Farmhouse vegetable soup
(serves 4)

Ingredients	
• 3 carrots • 1 potato • 1 parsnip • 1 leek • 1 onion • 3 fresh or 1 tin tomatoes • 8 mushrooms	• 1 teaspoon olive oil • 50 g flour • 150 ml milk • 1 litre vegetable stock (use two stock cubes) • salt and pepper *Garnish:* parsley, cream (optional)

Method
1. Wash, peel and dice carrots, potato and parsnip. Wash and chop leek and onion. Skin and chop tomatoes (if using fresh). Wash and slice mushrooms.
2. Heat the oil in a large heavy saucepan. Gently fry the onion and mushrooms.
3. Add carrots, potatoes and parsnips. Continue to fry gently.
4. Add flour and gradually stir in the milk.
5. Add the stock, tomatoes, salt and pepper. Bring to the boil. Once boiling, reduce the heat, cover and simmer for 45 minutes. Purée in a blender if you wish.
6. Serve piping hot in warm bowls. Garnish with a swirl of cream and some chopped parsley.

ESSENTIALS FOR LIVING

Work card 6: Workbook p. 113

Cream of mushroom soup
(serves 4)

Ingredients	
• 500 g mushrooms • 1 medium onion • 1 clove garlic • 1 dessertspoon olive oil • 25 g flour	• 150 ml milk • 500 ml vegetable stock • salt and pepper *Garnish:* cream and parsley

Method

1. Wash and slice mushrooms. Peel and chop onion. Peel and crush garlic.
2. Heat oil in a medium-sized heavy saucepan. Gently fry onion and garlic. Add mushrooms, keep frying gently for 3–5 minutes.
3. Add flour, salt and pepper. Gradually stir in the milk.
4. Add the stock, bring to the boil. Once boiling, reduce the heat and simmer for 30 minutes. Purée in a blender.
5. Serve piping hot in warm bowls. Garnish with a swirl of cream and chopped parsley.

Top Tip:
Serve puréed soups with crunchy croutons for more texture.

How to make croutons
Ingredients:
2 slices thick bread (cubed)
1 tablespoon olive oil
1 clove garlic (crushed)
¼ teaspoon salt
Method:
1. Toss bread cubes, salt, oil and garlic in a plastic freezer bag.
2. Bake for 15 minutes at 180°C / Gas 4.

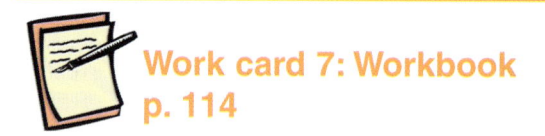

Work card 7: Workbook p. 114

Tomato and basil soup
(serves 4)

Ingredients	
• 1 medium onion • 1 clove garlic • 450 g fresh or tinned tomatoes • 1 teaspoon olive oil • 2 dessertspoons tomato purée • 500 ml vegetable stock • salt and pepper	• 2 dessertspoons chopped fresh basil or • 1 dessertspoon dried basil *Garnish:* fresh basil

128

RECIPES

Method
1. Peel and chop onion, peel and crush garlic. Skin and chop tomatoes (if using fresh).
2. Heat oil. Fry onion and garlic gently. Add tomatoes and tomato purée, stock, salt, pepper and basil.
3. Bring to the boil. Reduce the heat and simmer with the lid on for 30 minutes.
4. Purée soup and serve in warm bowls. Garnish with sprigs of fresh basil.

Work card 8: Workbook p. 115

SNACKS
Croque monsieur
(serves 1)

Ingredients	
• 25 g grated Cheddar cheese	• 1–2 slices cooked ham
• 2 slices bread	*Garnish:* parsley
• a little butter or low-fat spread	

Method
1. Grate the cheese (use large-hole side of grater).
2. Don't butter the bread. Make a layered sandwich – 1/3 cheese – ham – 1/3 cheese.
3. Butter top slice, place buttered side up. Grill until brown.
4. Turn over, sprinkle remaining cheese on top. Brown under grill. Cut into triangles.

5. Serve hot on a plate, garnish with fresh parsley.

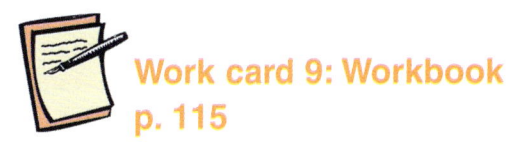

Work card 9: Workbook p. 115

French toast sandwiches
(serves 2)

Ingredients	
• 4 slices thick pan bread	• 75 g cherry tomatoes (halved)
• 2 eggs	• 50 g grated Cheddar cheese
• 4 tablespoons milk	*Garnish:* parsley
• a little oil	

Method
1. Trim the crusts off the bread. Beat the eggs and milk.
2. Pour the eggs into a large shallow dish. Lay the bread in this mixture for a few minutes – turn bread over and allow it to absorb the egg.
3. Heat the oil in a frying pan. Fry bread for one minute on each side. Remove bread, drain on kitchen roll.
4. Add tomatoes to the pan, toss gently for 30 seconds until heated.
5. Make a sandwich with the slices of French toast and the tomatoes.
6. Top each sandwich with grated cheese. Garnish with parsley. Serve immediately.

Work card 10: Workbook p. 116

ESSENTIALS FOR LIVING

French bread pizza
(serves 2)

Ingredients	
• 1 French stick • 4 tablespoons tomato ketchup • black pepper • 1 clove garlic, crushed • 1 teaspoon oregano	• 1 small onion, chopped • 2 tomatoes, sliced • 100 g grated Cheddar cheese or mozzarella

Method
1. Cut French stick in half and cut each piece in half again.
2. Mix ketchup, pepper, garlic and oregano. Spread over 4 slices.
3. Top with onion, tomato and cheese.
4. Place under a preheated grill until brown (5–7 minutes).
5. Serve with a tossed green salad.

Work card 11: Workbook p. 116

Main courses

BEEF, PORK AND LAMB

Spaghetti bolognese
(serves 4)

Ingredients	
• 1 onion • 1 clove garlic • 2 streaky rashers (chopped) • 100 g mushrooms • 1 tablespoon olive oil • 200 g lean minced beef • 1 tin chopped tomatoes or 4 ripe fresh tomatoes (skinned)	• 2 tablespoons tomato purée • 1 teaspoon dried oregano • salt and black pepper *To serve:* 200 g spaghetti, parsley, Parmesan cheese

Method
Bolognese sauce:
1. Peel and chop onion. Peel and crush garlic. Cut rashers into small pieces with clean scissors. Wash and slice mushrooms.

② Heat olive oil in a medium-sized saucepan. Fry onion, garlic, rashers and mushrooms gently for 5 minutes.
③ Add mince and fry until it has lost its red colour.
④ Add tomatoes, tomato purée, oregano, salt and pepper. Bring to the boil. Turn down the heat and simmer for 20–25 minutes.

To cook spaghetti:
⑤ Boil 3 litres of water in a large saucepan. Add 1 dessertspoon olive oil.
⑥ When water is boiling ease spaghetti into it. Boil for 10–15 minutes until spaghetti is 'al dente' – this means cooked but still has bite.

To serve:
⑦ Serve spaghetti on a warm plate. Spoon bolognese sauce on top. Garnish with parsley. Serve with Parmesan cheese and garlic bread (see page 127).

Work card 12: Workbook p. 116

Lasagne

Ingredients	
• 1 quantity of bolognese sauce (see recipe above) • 9 sheets lasagne *Cheese sauce:* • 750 ml milk • 50 g margarine	• 50 g flour • ½ teaspoon dry mustard • 150 g Cheddar cheese • salt and pepper *Garnish:* parsley

Method
① Make bolognese sauce as described for spaghetti bolognese.
② Grease a large rectangular pie dish.
③ Make the cheese sauce by whisking the milk, margarine, flour, salt, pepper and mustard in a saucepan and heat until it boils and thickens.
④ Add 100 g of the cheese. Leave 50 g aside.
⑤ Layer the meat sauce, lasagne sheets and cheese sauce as shown. Finish with a layer of cheese sauce. Sprinkle remaining cheese on top.
⑥ Bake at 190°C / Fan 180°C / Gas 5 for 30 minutes.
⑦ Garnish with parsley. Serve with garlic bread and a tossed green salad.

Work card 13: Workbook p. 117

Beef or lamb curry
(serves 4)

Ingredients	
• 500 g stewing beef or lamb • 1 dessertspoon olive oil • 1 large onion (chopped) • 1 clove garlic (crushed) • 25 g flour • 2 dessertspoons curry powder	• 1 litre stock (use two beef stock cubes) • 1 large cooking apple (peeled and chopped) *Note:* add no extra salt: plenty with stock cubes

Method
1. Remove fat from meat and cut into small pieces.
2. Heat olive oil in a medium-sized saucepan. Fry meat, chopped onion and garlic until meat is brown on all sides.
3. Add flour and curry powder, cook for 2–3 minutes.
4. Gradually stir in stock, add apple.
5. Bring to the boil, reduce heat and simmer for 1–1½ hours until meat is tender.
6. Serve on a bed of rice, garnish with parsley.

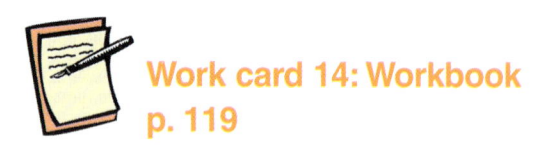

Work card 14: Workbook p. 119

Home-made burgers
(serves 4)

Ingredients	
• 450 g lean minced beef • 1 small onion (finely chopped) • 2 tablespoons wholemeal breadcrumbs • pinch of mixed herbs	• 1 egg (beaten) • salt and pepper • a little flour *To serve:* burger bun, lettuce, tomato, ketchup

Method
1. Mix mince, onion, breadcrumbs, herbs, pepper and salt in a bowl.
2. Bind together with beaten egg. Divide mixture into 4.
3. Shape on a floured board.
4. Brush each burger with a little olive oil.
5. Place burgers under a preheated hot grill for 5 minutes on each side.
6. Turn again and grill for a further 2 minutes each side.
7. Serve on a toasted burger bun with lettuce, tomato slices and ketchup.

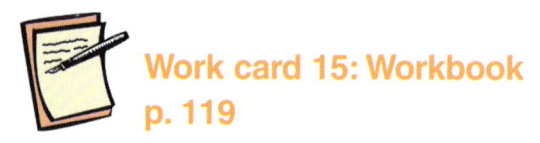

Work card 15: Workbook p. 119

RECIPES 14

Stuffed pork chops
(serves 4)

Ingredients	
• 4 lean pork chops • salt and pepper • olive oil • 8 mushrooms (sliced)	• 1 dessertspoon lemon juice • 1 dessertspoon flour • teaspoon mixed herbs

Method
Preheat oven to 160°C / Fan 150°C / Gas 3.
1. Trim excess fat off chops. Season on both sides with salt and pepper.
2. Fry chops in a little olive oil to seal in the juices. Remove from the pan.
3. Cook mushrooms in the pan for a few minutes, then add lemon juice, flour and herbs. Cook for a further 2–3 minutes. Remove from the heat.
4. Place each chop on a piece of tin foil large enough to wrap it in.
5. Pile some of the mushroom mixture onto each chop. Seal each chop in foil.
6. Place on a baking tray. Cook for 35–40 minutes.
7. Serve with potatoes and fresh vegetables.

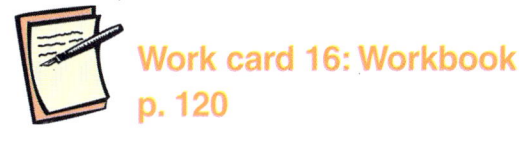

Work card 16: Workbook p. 120

Stuffed liver and bacon
(serves 2)

Ingredients	
• 25 g butter • ¼ onion (finely chopped) • 1 clove garlic (crushed) • 50 g breadcrumbs	• 1 teaspoon chopped parsley • 4 slices of lamb's liver • 4 streaky rashers • salt and pepper

Method
Preheat oven to 190°C / Fan 180°C / Gas 5.
1. Make the stuffing:
   - Melt butter in a small saucepan.
   - Add chopped onions, crushed garlic, breadcrumbs, salt, pepper and parsley.
2. Wash and dry liver. Place on a baking tray. Pile stuffing on top of each piece. Cover with streaky rasher. Secure with cocktail sticks.
3. Bake uncovered for 20 minutes. Cover with foil and bake for a further 15–20 minutes.
4. Serve on a warm plate with potatoes and fresh vegetables.

Work card 17: Workbook p. 120

ESSENTIALS FOR LIVING

CHICKEN

Fruity chicken curry
(serves 4)

Ingredients
• 4 chicken breasts • 1 dessertspoon flour • 1 dessertspoon olive oil • 250 ml chicken stock • 1 onion (chopped) • 1 tin plum tomatoes (drained) • 8 mushrooms • small tin pineapple chunks • 1 clove garlic (crushed) ***Serve with:*** • 2 dessertspoons medium curry powder 400 g cooked rice (300 g uncooked)

Method
1. Chop each chicken breast into bite-sized pieces.
2. Heat the oil in a medium-sized saucepan. Gently fry the onion, mushrooms and garlic. Add the chicken and fry until golden brown.
3. Add the curry powder and flour. Cook for 2–3 minutes. Keep stirring or mixture will stick.
4. Stir in stock, tomatoes and pineapple.
5. Bring to the boil. Turn down the heat and simmer for 40 minutes.
6. Serve on a bed of rice, garnish with parsley.

Work card 18: Workbook p. 120

Chicken and bacon risotto
(serves 4)

Ingredients
• 2 chicken breasts • 2 tablespoons olive oil • 3 streaky rashers • 6 mushrooms • 1 medium onion • 500 ml chicken stock • 1 clove garlic • salt and black pepper • 1 green pepper • 300 g uncooked rice • 1 red pepper

Method
1. Prepare ingredients:
   • Cut chicken into bite-sized chunks.
   • Cut rashers into medium-sized pieces with clean scissors.
   • Peel and chop onion.
   • Peel and crush garlic.
   • Wash, de-seed and chop peppers.
   • Wash and slice mushrooms.
2. Heat 1 tablespoon of oil in a large heavy saucepan. Fry rashers and onions until onions soften and rashers begin to brown.
3. Add garlic, mushrooms and peppers. Fry gently for another minute. Remove rashers and vegetables to a plate.

❹ Heat 1 tablespoon of oil in the saucepan. Fry chicken for 3 minutes. Add rice, fry for another minute. Add salt and pepper.

❺ Return rashers and vegetables with stock to the saucepan. Bring to the boil.

❻ Once boiling, reduce heat and simmer for 25 minutes. Stir now and then to prevent sticking.

❼ Serve on a warm oval plate. Garnish with parsley.

Work card 19: Workbook p. 122

Sweet and sour chicken or pork

Ingredients	
• 3 pork chops or • 2 chicken breasts • 1 onion • 1 clove garlic • 1 green pepper • 1 carrot • 1 tablespoon olive oil	• ½ can (400 g) pineapple in own juice • 1 dessertspoon cornflour • 1 tablespoon vinegar • 1 tablespoon soy sauce • 1 teaspoon brown sugar • 1 teaspoon tomato purée • salt and black pepper

Method

❶ Prepare ingredients:
- Wipe meat, trim fat and cut into pieces.
- Peel and chop onion.
- Peel and crush garlic.
- Wash, de-seed and dice pepper.
- Wash, peel and slice carrot into strips.

❷ Heat the oil in a medium-sized heavy saucepan. Fry the meat quickly to seal it. Remove meat onto a plate.

❸ Sauté onion, garlic, carrot and pepper for 2–3 minutes. Return meat.

❹ Measure the pineapple juice, make it up to 400 ml with water. Blend cornflour with a little of this juice.

❺ Add pineapple juice, blended cornflour and all other ingredients to the saucepan.

❻ Bring to the boil, turn down the heat and simmer with the lid on for 30 minutes.

❼ Serve on a bed of rice, garnish with parsley.

Work card 20: Workbook p. 122

FISH

Tomato and fish bake

Ingredients	
• 700 g cod or haddock (filleted and skinned) *Tomato sauce:* • 1 small onion • 1 clove garlic • 2 sticks celery • 1 teaspoon olive oil • 1 tin chopped tomatoes	• 1 teaspoon sugar • 1 teaspoon dried basil • salt and black pepper *Topping:* • 1 teaspoon vegetable oil • 50 g brown breadcrumbs • 25 g grated Cheddar cheese

Method

Preheat oven to 200°C / Fan 190°C / Gas 6.

1. Wash, dry, fillet and skin the fish (see pages 73–74). Cut fish into bite-sized pieces. Place on a lightly greased oven-proof dish.
2. Prepare ingredients for the tomato sauce:
   - Skin and chop onion finely.
   - Peel and crush garlic.
   - Wash and slice celery.
3. Make tomato sauce:
   - Heat oil in a small saucepan.
   - Gently fry the onion, garlic and celery for 2–3 minutes.
   - Add tomatoes, sugar, basil, salt and black pepper.
   - Bring to the boil, reduce heat and simmer for 10 minutes – stir occasionally.
4. Make topping:
   - Heat olive oil in saucepan.
   - Add breadcrumbs and then mix in grated cheese.
5. Pour tomato sauce over the fish. Sprinkle the topping over.
6. Bake for 30 minutes until topping is crisp and brown.

Work card 21: Workbook p. 123

Fish cakes

(serves 4)

Ingredients	
• 350 g cooked white fish • 6 potatoes (cooked) • 2 eggs • 2 dessertspoons parsley	• 25 g butter • 75 g breadcrumbs (3 slices) • olive oil • 50 g flour • salt and black pepper

Method

1. Prepare ingredients:
   - Flake fish with a fork.
   - Mash potatoes.
   - Beat one egg.
   - Chop parsley.
   - Melt butter.

❷ Mix fish, potatoes, egg, parsley, butter, salt and pepper in a bowl. Leave in the fridge for 30 minutes.

❸ Remove from the fridge and roll into a long 'snake' shape on a floured board. Cut 'snake' into 8 even-sized pieces.

❹ Flatten each piece into a round cake. Beat the second egg. Dip each cake in egg and then in breadcrumbs.

❺ Fry or grill cakes (grilling is healthier) until golden brown on each side. Serve with fresh vegetables.

Top Tip:
In a hurry? Use tinned salmon or tuna.

Work card 22: Workbook p. 123

MILK, EGGS AND CHEESE

Quiche Lorraine
(serves 4)

Ingredients	
Pastry: (short-crust) • 100 g flour • 50 g margarine • pinch salt • cold water	*Filling* • 4 back rashers • 50 g grated cheese • 2 eggs • 200 ml milk • salt and pepper

Method
Preheat oven to 200°C / Fan 190°C / Gas 6.

❶ Make pastry:
  • Sieve flour and salt into a bowl.
  • Rub in chopped margarine until it looks like breadcrumbs.
  • Add water little by little with a spoon. Stir with a knife until you have a stiff dough.
  • Roll out and line a 20 cm flan tin.

❷ Grill the rashers and chop them up. Place them in the flan tin with half of the grated cheese.

❸ Mix together the eggs, milk, pepper and salt. Pour over the rashers. Sprinkle the remaining cheese on top.

❹ Bake for 40–45 minutes until set.

❺ Garnish with tomato slices and parsley. Serve with a side salad.

Top Tip:
Fillings can be varied. Try adding 100 g mushrooms: grill the mushrooms, allow to cool, slice and arrange on flan tin with rasher and cheese; continue as above.

Work card 23: Workbook p. 124

ESSENTIALS FOR LIVING

Pizza
(serves 4)

Ingredients	
Scone base: • 200 g flour • 50 g margarine • 1 egg • a little milk • pinch of salt	*Topping:* • ½ onion • 4 mushrooms • 1 tin tomatoes or 6 fresh tomatoes (skinned) • 2 dessertspoons tomato purée (if using fresh tomatoes) • 1 dessertspoon olive oil • 1 teaspoon oregano or basil • 75 g grated cheese

Method
Preheat oven to 190°C / Fan 180°C / Gas 6.

❶ Make scone base:
- Sieve flour and salt into a bowl.
- Rub in chopped margarine until it looks like breadcrumbs.
- Add egg and enough milk to make a stiff dough.
- Roll out into a circle. Place on a greased tin.

❷ Make topping:
- Chop onion, mushrooms and tomatoes.
- Heat the oil in a small saucepan. Fry the onion and mushrooms gently for 2–3 minutes.
- Add the tomatoes, purée and herbs.
- Simmer gently for 5 minutes with the lid off.

❸ Spread tomato mixture on the base. Sprinkle cheese on top.

❹ Bake for 20–25 minutes until cheese is bubbling.

Top Tip:
Other toppings: cooked ham, pineapple chunks, red/green pepper, pepperoni slices. Mozzarella cheese can also be used.

Work card 24: Workbook p. 125

VEGETARIAN

Vegetable stir-fry with rice
(serves 4)

Ingredients	
• 400 g cooked rice (300 g uncooked) *Stir-fry:* • ½ onion • 1 clove garlic • 1 carrot • 1 red pepper • 1 head broccoli • 5–6 cauliflower florets • 5 mushrooms	• 1 small courgette • 50 g bean sprouts • 50 g green beans • 1 tablespoon olive oil • 1 teaspoon cornflour • 2 tablespoons soy sauce • 50 g peanuts or cashew nuts

RECIPES 14

Note: some of the vegetables above are optional, e.g. bean sprouts, green beans, cauliflower florets.

Vegetable stir-fry

Method

1. Put the rice on to cook in a saucepan of boiling water to which you have added one dessertspoon of olive oil.
2. Prepare all vegetables for stir-frying:
   - Peel and chop onion.
   - Peel and crush garlic.
   - Peel and cut carrot into thin batons.
   - Wash, de-seed and slice pepper.
   - Wash and break broccoli and cauliflower into small florets.
   - Wash and slice mushrooms and courgette.
   - Wash green beans and sprouts.
3. Mix all the vegetables together in a bowl.
4. Heat oil in a wok or frying pan. Add the vegetables and stir well.
5. Add the cornflour, stir through, add the soy sauce and nuts. Continue stir-frying for 10–15 minutes until vegetables are cooked but still crisp. Serve with boiled rice.

Work card 25: Workbook p. 125

Vegetable curry
(serves 4)

Ingredients	
1 kg mixed vegetables: • onion (1) • carrot (1) • small courgette (1) • mushrooms (8) • celery (2 sticks) • cauliflower (6 florets) • broccoli (6 florets) • green or red pepper (1) • 1 cooking apple	• 500 ml vegetable stock • 1 dessertspoon olive oil • 1 dessertspoon flour • 3 dessertspoons curry powder • 1 dessertspoon coconut • 50 g sultanas • 1 teaspoon lemon juice • 1 teaspoon brown sugar • ½ tin pineapple chunks (drained) • salt and pepper

Method

1. Prepare vegetables:
   - Peel and chop onion.
   - Peel and slice carrot.
   - Wash and slice courgette, mushrooms and celery.
   - Wash and divide cauliflower and broccoli florets into bite-sized pieces.
   - Wash, de-seed and slice pepper.
   - Peel, core and chop apple.
2. Make vegetable stock.
3. Heat the olive oil in a small saucepan. Fry

139

the onion gently for 2–3 minutes. Add the flour and curry powder, gradually stir in 100 ml of the stock. This is the curry sauce.

❹ Put the rest of the ingredients into a large saucepan. Add 400 ml of stock. Bring to the boil, reduce the heat and simmer with the lid on for 15 minutes.

❺ Add the curry sauce to the large saucepan. Bring to the boil again, reduce the heat and simmer for a further 10–15 minutes.

❻ Serve on a bed of rice, garnish with parsley.

Work card 26: Workbook p. 126

Cheesy vegetable and pasta bake
(serves 4)

Ingredients	
Vegetable mix: • 1 dessertspoon olive oil • 1 onion • 1 clove garlic • 1 green pepper • 8 mushrooms • 1 tin chopped tomatoes • 1 teaspooon dried basil *Pasta:* • 200 g pasta shapes • 1 dessertspoon olive oil • boiling water	*Cheese sauce:* • 25 g margarine • 25 g flour • 500 ml milk • 50 g grated cheese *Topping:* • 50 g brown breadcrumbs • 25 g grated Cheddar cheese *Garnish:* parsley

Method
Preheat oven to 200°C / Fan 190°C / Gas 6.

❶ Heat oil in a medium-sized saucepan. Fry the onion, garlic, green pepper and mushrooms gently for 2–3 minutes.

❷ Add the tomatoes and basil. Bring to the boil. Reduce the heat and simmer with the lid on for 20 minutes.

❸ Cook the pasta shapes in boiling water, to which oil has been added, for 12–15 minutes. Drain and add to the vegetable mix.

❹ Make cheese sauce:
- Melt the margarine in a small saucepan.
- Add the flour; cook this roux for 1 minute stirring all the time.
- Remove from heat, cool slightly.
- Add the milk bit by bit stirring all the time. Return to heat and bring to the boil. Reduce heat and cook for 5 minutes.
- Remove from the heat, stir in the grated cheese.

❺ Put the vegetables and pasta in a casserole dish. Next pour on the cheese sauce and then top with the grated cheese and breadcrumb mix.

❻ Bake for 15 minutes. Serve hot with garlic bread, garnish with parsley.

Work card 27: Workbook p. 127

DESSERTS

Fresh fruit salad

(serves 4)

Ingredients	
• 1 apple • 1 pear • 1 orange • 6 green seedless grapes • 6 black seedless grapes • 6 strawberries	• 2 kiwi fruit • 1 banana • 200 ml orange juice *To serve:* • low-fat cream, ice cream or yogurt

Method

1. Prepare the fruit:
   - Core and cut apple and pear into bite-sized pieces.
   - Remove pith and skin from orange. Break into segments, cut each segment in two.
   - Halve grapes and strawberries.
   - Peel and slice kiwi and banana.
2. Arrange fruit in a bowl. Pour orange juice over.
3. Serve with low-fat ice cream, cream or natural yogurt (healthier option).

Top Tip:
Orange juice is a healthy alternative to sugar syrup. It also prevents apples, pears and bananas from browning.

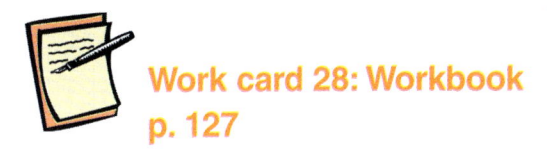

Work card 28: Workbook p. 127

Apple and plum crumble

(serves 4)

Ingredients	
• 2 cooking apples • 50 g sugar • 3 ripe plums	*Crumble:* • 150 g flour • 50 g margarine • 50 g brown sugar • teaspoon cinnamon

Method

Preheat oven to 180°C / Fan 170°C / Gas 5.
1. Put peeled sliced apples and sugar in a saucepan, cook gently until soft. Place in a pie dish.
2. Peel and de-stone plums. Slice and layer on top of apple.
3. Make crumble:
   - Sieve flour.
   - Rub in chopped margarine until it looks like breadcrumbs.
   - Add brown sugar and cinnamon.
4. Spread crumble evenly over fruit.
5. Bake for 30 minutes until golden on top.
6. Serve with low-fat ice cream, cream or natural yogurt (healthier option).

> *Top Tip:*
> Many different fruits can be used to make fruit crumble e.g.
> - 6 stalks of rhubarb
> - one tin pears or three ripe fresh pears
> - 200 g blackberries
> - 200 g gooseberries.

Work card 29: Workbook p. 128

Bread and butter pudding
(serves 4)

Ingredients	
• 8 slices of bread	• 300 ml milk
• 50 g butter	• vanilla essence
• 50 g sultanas	• 25 g sugar
• nutmeg	• 2 eggs

Method
Preheat oven to 190°C / Fan 180°C / Gas 5.
1. Lightly butter the bread, remove black crusts only. Cut 6 slices into fingers. Cut remaining slices into triangles.
2. Line the bottom of an oven-proof dish with bread, butter side down. Sprinkle with sultanas and a little nutmeg.
3. Repeat until all the bread is used, finishing with a layer of triangular-shaped bread, butter side up.
4. In a saucepan heat the milk, vanilla essence and sugar until hot – not boiling.
5. Beat eggs in a bowl, add milk to eggs gradually, keep stirring.
6. Pour this mixture over the bread and sultanas. Sprinkle with nutmeg and a little sugar.
7. Place dish on a roasting tin. Put some water in the roasting tin: this helps keep the pudding moist.
8. Bake for 30–35 minutes. Serve with low-fat ice cream, cream or natural yogurt (healthier option).

> *Top Tip:*
> Bread and butter pudding can be made in small individual ramekin dishes (grease ramekin dishes lightly). Tip each one out upside down on a dessert plate. Serve surrounded with hot custard.

Work card 30: Workbook p. 128

Delicious filled pancakes
(makes six pancakes)

Ingredients	
Pancakes:	*Fillings (pick one):*
• 100 g flour	• ice cream, banana and butterscotch sauce
• 1 egg	
• pinch salt	• strawberries and whipped cream
• 250 ml milk	
• oil for frying	• cinnamon-flavoured stewed apple

Method

1. Sieve the flour and salt into a bowl. Make a well in the centre.
2. Drop egg and half the milk into well. Using a whisk beat from the centre out until the mixture is smooth.
3. Add the rest of the milk. Beat for 5 minutes to introduce air.
4. Pour into a jug. Allow to stand in the fridge for 20 minutes if you have time.
5. Heat a little oil in a frying pan. Pour some batter mixture and tilt to cover the base. Cook until edges begin to brown. Shake the pan to loosen pancake.
6. Turn the pancake using fish slice or palette knife. Repeat with rest of batter. Keep cooked pancakes warm between two plates over a saucepan of boiling water.
7. Serve 2 pancakes per person on a dessert plate.

Fillings:

Ice cream, banana with butterscotch sauce: fill pancakes with scoop of ice cream and ½ a banana (sliced), drizzle butterscotch sauce over.

Strawberry or apple: fill with 4 strawberries (sliced) or 1 dessertspoon of apple (stewed with cinnamon). Fold over and serve with whipped cream. Dredge with icing sugar.

Work card 31: Workbook p. 129

Rich chocolate orange sauce

Ingredients
• 100 g plain cooking chocolate
• zest of 1 orange
• juice of ½ orange
• 4 tablespoons whipping or double cream

Method

1. Melt chocolate carefully in a microwave or over a saucepan of barely simmering water.
2. Mix in zest, juice and cream.
3. Serve.

Top Tip:
This sauce is *very* high in calories – serve only very occasionally.

Butterscotch sauce

Ingredients	
• 50 g butter	• 100 ml double cream
• 75 g brown sugar	• ½ teaspoon vanilla essence
• 50 g table sugar	
• 4 dessertspoons golden syrup	

Method

1. Put the butter, sugars and syrup into a heavy saucepan. Melt on a low heat. Simmer for 3–5 minutes stirring all the time.
2. Remove from the heat, gradually stir in the cream and vanilla essence. Return to the heat for 2 minutes.

ESSENTIALS FOR LIVING

> *Top Tip:*
> This sauce is very high in calories – serve only occasionally.

Fruit coulis

Ingredients	
• 100 g raspberries (fresh, frozen or tinned) • 100 g strawberries (fresh, frozen or tinned)	• 1 tablespoon lemon juice • 1 tablespoon icing sugar

Method
1. Wash fruit. Put all the ingredients in a food processor or liquidiser. Blend until smooth.
2. Put the mixture through a sieve to remove seeds.

A fruit coulis can be served with almost any dessert. Examples: rice pudding, ice cream, bread and butter pudding, pancakes etc.

PASTRY

Short-crust pastry

Ingredients
• 200 g plain flour • 100 g margarine • cold water • pinch salt

Method
1. Sieve flour and salt into a bowl.
2. Cut the margarine into pieces, add it to the flour. Rub the margarine into the flour with the fingertips until it looks like breadcrumbs.
3. Add water with a spoon. Mix with a knife until pastry comes together in a ball. *Do not make it too wet.*
4. Leave to relax in the fridge until needed.

> *Top Tip:*
> Do not handle pastry too much; keep it as cool as possible.

Variations for savoury dishes, e.g. quiche

Wholemeal pastry: Rich in fibre. Use ½ wholemeal flour, ½ plain flour.

Cheese pastry: Add 75 g grated cheese and ¼ teaspoon of mustard to basic short-crust pastry recipe (add before water).

> *Top Tip:*
> If a recipe says 200 g of pastry this refers to the amount of flour in the recipe.

RECIPES 14

Sausage rolls

(makes 10 sausage rolls, 20 mini-rolls)

Ingredients
• 200 g short-crust pastry • 200 g sausage meat • beaten egg to glaze

Method

Preheat oven to 200°C / Fan 190°C / Gas 6.

1. Make short-crust pastry (see page 144).
2. Roll out pastry into a rectangle slightly bigger than this page. Divide into 2.
3. Divide the sausage meat into 2. On a slightly floured board roll each into a long sausage shape the length of the pastry.

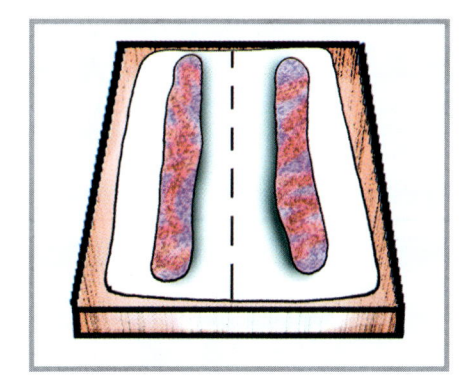

4. Place sausage meat on the pastry and dampen the edges.
5. Roll the pastry around the sausage meat. Brush with egg. Cut each roll into 5 smaller rolls (cut into 10 for mini-rolls). Make 2–3 slits on top of each one.
6. Bake for 10 minutes at 200°C / Fan 190°C / Gas 6. Reduce the heat a little and cook for a further 10–15 minutes.
7. Garnish with parsley.

Work card 32: Workbook p. 130

Mince pies

Ingredients
• 200 g short-crust pastry • 150 g mincemeat • egg to glaze • icing sugar to decorate

Method

Preheat oven to 200°C / Fan 190°C / Gas 6.

1. Make the short-crust pastry (see page 144).
2. Roll out 2/3 of the pastry thinly. Cut out 12 circles using large (6 cm) cutter. Place each in patty tin. Dampen edges with water.
3. Put one teaspoon of mincemeat in each.
4. Roll out the rest of the pastry. Use smaller cutter (5 cm) to cut out 12 lids. Put lid on each pie. Seal with fingers.
5. Glaze with egg.
6. Bake at high temperature for 10 minutes. Reduce the heat a little and cook for a further 10 minutes.
7. To decorate, sprinkle icing sugar over pies through a sieve.
8. Place on a round plate with a doily. Serve with whipped cream or brandy butter.

> *Top Tip:*
> When rolling out pastry, roll lightly in one direction only. Keep turning the pastry to stop it sticking to the table.

Work card 33: Workbook p. 130

BREAD AND CAKES

Rubbing-in method
- brown bread
- scones
- banana bread
- coconut buns
- rock buns

Brown bread

Ingredients	
• 150 g white flour	• 25 g bran
• 1 level teaspoon bread soda	• 25 g wheat germ (optional)
• 150 g wholemeal flour	• 1 egg
• 50 g margarine	• 350 ml approx. buttermilk
	• 25 g castor sugar

Method
Preheat oven to 190°C / Fan 180°C / Gas 5.
1. Grease a 1 lb loaf tin.
2. Sieve white flour and bread soda into a bowl. Mix in wholemeal flour.
3. Rub in margarine. Add bran, wheat germ and sugar.
4. Add egg and buttermilk. Mix to a soft wet dough.
5. Spread into loaf tin. Cook for approximately one hour.
6. Turn onto a wire tray. Allow to cool before cutting.

> *Top Tip:*
> Tap the base of the loaf: if it sounds hollow it's cooked.

Work card 34: Workbook p. 131

Tea scones

Ingredients	
• 200 g self-raising flour	• 1 egg
• pinch salt	• a little milk
• 50 g margarine	• egg to glaze
• 25 g castor sugar	

Method
Preheat oven to 220°C / Fan 210°C / Gas 7.
1. Sieve the flour and salt into a bowl. Rub in the margarine until the mixture looks like breadcrumbs.
2. Add the sugar. Beat the egg and milk in a bowl. Add to the mixture.
3. Turn the mixture out onto a floured board and knead lightly.

RECIPES 14

④ Roll out, but not too thinly (see Top Tip). Cut out into 8–10 rounds.
⑤ Place scones onto greased baking tin. Glaze with egg.
⑥ Bake for 15 minutes until golden. Cool on a wire tray.

Variations:

Fruit scones: Add 25 g raisins with or instead of sugar.
Brown scones: Use 100 g of brown flour.

Top Tip:
Don't roll out scone mixture too thinly: it should be 1.5 cm thick.

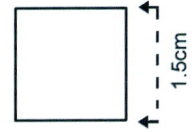

Work card 35: Workbook p. 131

Banana bread

Ingredients	
• 100 g plain flour	• 3 large ripe bananas
• 1 teaspoon mixed spice	• 1 egg
• 1 teaspoon baking powder	• 2 tablespoons honey
• 100 g wholemeal flour	• 100 ml oil
• 25 g walnuts chopped roughly (optional)	

Method
Preheat oven to 180°C / Fan 170°C / Gas 4.
① Grease a 1 lb loaf tin. Line the bottom with greaseproof paper.
② Sieve the plain flour, mixed spice and baking powder into a bowl. Add the wholemeal flour and walnuts, if using.
③ Mash up the ripe bananas. Add these, the oil, egg and honey to the flour. Mix well.
④ Spread into the loaf tin. Bake for 45 minutes.

Coconut buns

Ingredients	
• 200 g flour	• 50 g coconut
• pinch salt	• 1 egg
• 1 teaspoon baking powder	• a little milk
• 50 g margarine	*To decorate:*
• 50 g castor sugar	• 2 tablespoons jam
	• 2 heaped tablespoons coconut

Method

Preheat oven to 200°C / Fan 190°C / Gas 6.

① Sieve the flour, salt and baking powder into a bowl.
② Rub in the chopped margarine. Add sugar and coconut.
③ Add the egg and enough milk to make a stiff dough.
④ Using a spoon and a fork, pile the mixture into 10 heaps on a greased baking tray.
⑤ Bake for 20 minutes.
⑥ To decorate: beat the jam in a small bowl to make it soft. Put coconut onto a plate. Dip buns in jam and then in coconut.

> *Top Tip*:
> Never put tins or dishes on the floor of the oven: the bottom of the food will burn.

Work card 36: Workbook p. 132

Rock buns

Ingredients	
• 200 g flour	• 50 g dried fruit
• pinch salt	• 25 g mixed peel
• 50 g margarine	• 1 egg
• ½ teaspoon mixed spice	• a little milk
• 25 g castor sugar	• table sugar to sprinkle on top

Method

Preheat oven to 200°C / Fan 190°C / Gas 6.

① Grease patty tin.
② Sieve flour and salt into a bowl. Rub in margarine until it looks like breadcrumbs.
③ Add mixed spice, castor sugar, dried fruit and mixed peel. Mix well.
④ Add egg and enough milk to make a stiff dough.
⑤ Pile into patty tin. Mixture should make 8–10 cakes.
⑥ Sprinkle each cake with a little table sugar.
⑦ Bake for 20–25 minutes until golden. Cool on a wire tray.

Work card 37: Workbook p. 132

Creaming method

Queen cakes

Ingredients	
• 100 g castor sugar	• 150 g self-raising flour (or plain flour with 1 level teaspoon baking powder added)
• 100 g margarine	
• 2 eggs	
• 3 drops vanilla essence	
	• a little milk

Method

Preheat oven to 200°C / Fan 190°C / Gas 6.

① Beat sugar and margarine until it is white and creamy.
② Add eggs and vanilla essence a little at a time. Beat well between additions.
③ Gently fold in the sieved flour. If the mixture is too dry add a little milk.
④ Spoon into bun cases in a patty tin. Bake for 15–20 minutes until golden.

RECIPES 14

Top Tip:
A little flour can be added with the egg to stop the egg curdling. To test if queen cakes are cooked, check top and bottom of cake: both should be brown.

Work card 38: Workbook p. 132

Welsh cheese cakes

Ingredients	
Pastry: • 100 g flour • pinch salt • 50 g margarine • cold water *Filling:* • 1–2 tablespoons jam	*Madeira mix:* • 75 g flour • 50 g castor sugar • 50 g margarine • 1 egg • 1 drop vanilla essence

Welsh cheese cakes

Method

Preheat oven to 200°C / Fan 190°C / Gas 6.

1. Make short-crust pastry (see recipe on page 144).
2. Roll out the pastry thinly. Cut into 10–12 circles and place them in a greased patty tin.
3. Put a small amount of jam on each.
4. Make the Madeira mixture (as for Queen Cakes). Pile one large teaspoon of the mix on top of the jam in the patty tin. Smooth out with a wet knife to seal jam in.
5. Roll out pastry scraps. Cut thin strips. Lay on top of Madeira mix in a cross shape (see diagram).
6. Bake for approximately 15 minutes.
7. Cool on a wire tray.

Pastry strips

Work card 39: Workbook p. 133

All-in-one method

All-in-one chocolate cake

Ingredients	
• 175 g soft margarine • 175 g self-raising flour • 175 g castor sugar	• 3 eggs • 2 tablespoons drinking chocolate • 1 tablespoon boiling water

149

ESSENTIALS FOR LIVING

Method

Preheat oven to 180°C / Fan 170°C / Gas 4.
1. Grease two 18 cm sandwich tins. Line the bottom of each with circles of greaseproof paper.
2. Place all ingredients in a processor or bowl and beat until smooth.
3. Divide mixture between sandwich tins.
4. Bake for 25–30 minutes. Cool on a wire tray.
5. To decorate:
   - sandwich 2 cakes together with chocolate icing (see below)
   - spread icing on top of the cake
   - score the top of the cake with a fork or pipe roses of icing around edges.

Chocolate butter icing

Ingredients
• 100 g icing sugar
• 50 g soft margarine
• 1 tablespoon drinking chocolate
• 1 tablespoon milk

Method

Beat all the ingredients together in a bowl until completely smooth.

Work card 40: Workbook p. 134

Simple carrot cake

Ingredients	
• 225 g self-raising flour	*Topping:*
• 200 g grated carrot	• 50 g icing sugar
• 200 g castor sugar	• 100 g cream cheese (e.g. Philadelphia)
• 100 ml sunflower oil	• 3 drops vanilla essence
• 2 eggs	• walnuts to decorate

Method

Preheat oven to 190°C / Fan 180°C / Gas 4.
1. Line the bottom of 1 lb loaf tin with a rectangle of greaseproof paper.
2. Mix flour, carrots, sugar, oil and eggs together. Pile into the tin.
3. Bake for 60–70 minutes. Cool on a wire tray before decorating.

RECIPES **14**

④ Topping: cream sieved icing sugar, cheese and vanilla essence until really smooth. Spread onto the cake. Make lines in it with a fork. Decorate with pieces of walnut.

Work card 41: Workbook p. 134

Whisking method

Fresh strawberry sponge cake

Ingredients	
• 3 eggs • 75 g castor sugar • 75 g flour • ¼ teaspoon baking powder	• 125 ml whipped cream • 6 strawberries (sliced)

Method
Preheat oven to 190°C / Fan 180°C / Gas 5.
① Grease two 18 cm sandwich tins. Dust with a mix of castor sugar and flour.
② Whisk eggs and sugar until thick and creamy.
③ Fold half of the sieved flour and baking powder into the mixture, then fold in the other half.
④ Divide between 2 tins, bake for 15–20 minutes (see Top Tips).
⑤ Cool on a wire tray. Fill with whipped cream and sliced strawberries.

Top Tips:
1. If whisking egg and sugar by hand, place the bowl over a saucepan of hot water: mixture whisks more easily.
2. To check if a sponge is done, press on it. If it springs back, it is cooked.

Work card 42: Workbook p. 135

Melting method

Ginger snaps
(makes 12–15)

Ingredients
• 100 g self-raising flour • 1 heaped teaspoon ground ginger • 50 g margarine • 50 g table sugar • 3 dessertspoons golden syrup

Method
Preheat oven to 180°C / Fan 170°C / Gas 4.
① Sieve flour and ginger into a bowl.
② Gently melt margarine, sugar and syrup in a saucepan. Do not boil. Mix into flour.

❸ Make 12–15 balls of dough.
❹ Flatten into biscuit shapes on a greased tin.
❺ Bake for 15 minutes. Cool on the tin.

Work card 43: Workbook p. 135

Gingerbread
(makes 24 squares)

Ingredients	
• 100 g margarine • 50 g brown sugar • 3 tablespoons treacle • 150 g plain flour • ½ teaspoon mixed spice	• 1 heaped teaspoon ground ginger • ½ teaspoon baking powder • 1 egg • 100 g stewed apple

Method
Preheat oven to 180°C / Fan 170°C / Gas 4.
❶ Grease a large baking tray. Line the bottom with greaseproof paper.
❷ Gently melt margarine, sugar and treacle in a saucepan.
❸ Sieve flour, spice, ginger and baking powder into a bowl. Add melted ingredients and then beaten egg and apple.
❹ Pour the mixture into a baking tray. Bake for 15 minutes.
❺ Cut into 24 squares. Remove from tin and allow to cool on a wire tray.

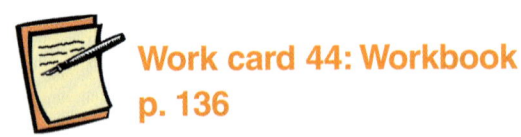

Work card 44: Workbook p. 136

Unit 2

Consumer Studies

chapter 15

What is a consumer?

'A consumer is someone who buys or uses goods and services.'

Examples of goods and services

Goods: things

- food
- drinks
- clothes
- toiletries e.g. shampoo
- CDs
- books
- computer games
- mobile phones
- sweets
- cigarettes
- heating
- lighting

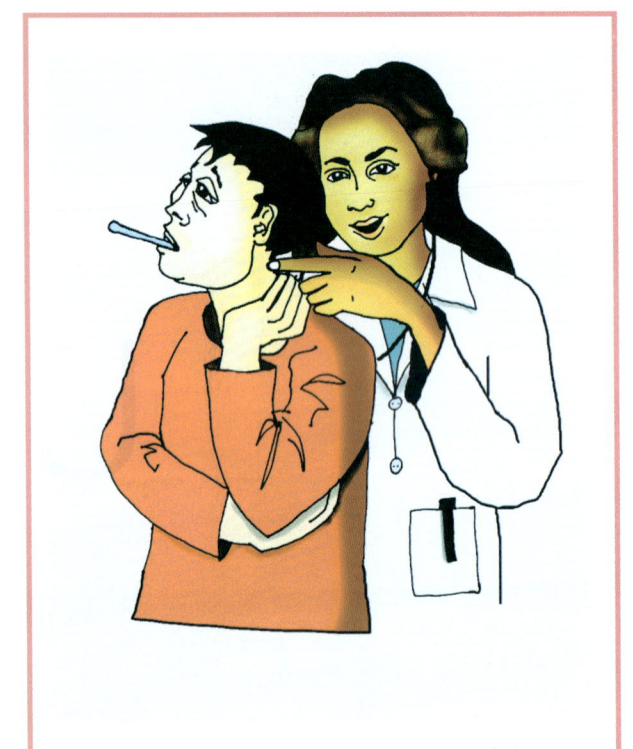

Services: people do something for you

- doctors
- dentists
- hairdressers
- refuse collection
- postal services
- public transport
- education
- taxis
- gardaí
- roads
- public parks
- street lighting

WHAT IS A CONSUMER? 15

Activity 15.1 – Workbook p 139

NEEDS AND WANTS

Needs are *essential* goods and services. Examples of needs are food, clothes, shelter, doctors and dentists.

Wants are *non-essential* goods and services. Examples are takeaways, cars, designer clothes, jewellery, beauticians and hairdressers.

Nowadays the line between what is a need and what is a want has become blurred. For example, many people would consider having a car a need, yet forty years ago very few people had cars. Today, in developing countries, a car is most definitely a want.

In a family where money is tight, debts build up when wants are bought before needs are paid for. For example buying a wide screen TV (non-essential) when the electricity bill (essential) hasn't been paid, or spending money on alcohol and cigarettes (non-essential) when there is no food (essential) in the house.

Factors that influence consumer decisions

- how much money is available
- peer pressure: you want what your friends have
- fashion trends
- advertising: advertisements try to make us feel 'if I have this product my life will be better in some way'
- personal values, for example a golfer who spends €650 on one golf club may be thought of as mad by non-golfers

Making wise consumer decisions

When we buy things on the spur of the moment, without thinking, this is called *impulse buying*. Goods and services bought in this way are often bad buys and ones we often regret. This regret is called *buyer's remorse*. When buying goods and services it is better to think carefully before we buy. Below are some of the factors that should be considered.

Factors to be considered when choosing a product

1. Money: How much can you afford?
2. Quality: Is the product made from good quality materials? Does it have any quality symbols on it?
3. Value: Shop around, compare similar products in other shops.
4. Suitability: Will it do the job you want it for? Example: hill-walking boots need to be waterproof.
5. Durability: Will it last? Examples: carpets, shoes.
6. Design: Is it good to look at, well finished?
7. Safety: Look for safety symbols, especially on children's toys and electrical goods.
8. Brand name: It is often safer to buy well known reliable brands. Example: washing machines.
9. Environmental impact: Is the product environmentally friendly?
10. After-sales service: Will the shop fix the product if it breaks down?
11. Maintenance: Is the product easy to keep clean?

ESSENTIALS FOR LIVING

12. Size, comfort: Are goods such as clothes and shoes well fitting and comfortable?

Sources of consumer information

Sometimes when we buy particular goods or services we are asked where we heard about them. Below are common sources of consumer information about goods and services.

- newspaper articles
- magazines
- golden pages
- advertisements
- word of mouth
- manufacturers' brochures and leaflets
- in the shop: sales people inform us; we also get information about products by looking at them ourselves

Golden Pages advertisement

Exam time – Workbook p. 139

Now test yourself at *www.my-etest.com*

chapter 16

Consumer protection

CONSUMER RIGHTS

We as consumers have five basic rights. A right is something you are entitled to.

The right to choice
This means having a wide variety of goods and services to choose from. Having choice means that manufacturers must produce high quality goods or they will not sell.

The right to information
This means that information written on goods or about services must be true. For example if a chair is labelled 'genuine leather' it must be leather.

The right to redress
If goods or services are faulty you are entitled to one of the following:
- Repair (item is fixed free of charge)
- Replacement (e.g. new pair of shoes)
- Refund (money back)

The right to quality
Goods must be able to do what they are meant to do, and be of good quality. Manufacturers usually test goods for quality before they leave the factory; this is called *quality control*.

The right to safety
Goods and services should be safe for consumers to use. Electrical items, children's clothes and toys must follow strict safety rules. Dangerous goods e.g. bleach must carry warning symbols.

ESSENTIALS FOR LIVING

> **A monopoly**
> This is when only one company provides a good or service, e.g. ESB, Iarnrod Eireann. What is the disadvantage of monopolies?

CONSUMER RESPONSIBILITIES

Above are listed your rights as a consumer. As a consumer you also have *responsibilities*.

You should:
- be well informed about the products you are buying and your consumer rights
- examine products and services carefully before you buy
- read instructions and labels; heed warnings
- shop around for value for money and avoid unnecessary waste

CONSUMER LAW

Consumers are protected in three ways:
1. The law
2. Government organisations, e.g. the ombudsman and the Director of Consumer Affairs (Chapter 17)
3. Voluntary organisations, e.g. Consumers' Association of Ireland (Chapter 17)

Contract

When you buy something in a shop or pay for a service a *legal contract* is formed between you (the buyer) and the seller, for example the shop-keeper or plumber. The contract *is not* between you and the manufacturer of the goods, for example Levi Strauss & Co.

The two most important consumer laws are:
- The Sale of Goods and Supply of Services Act, 1980
- The Consumer Information Act, 1978

The Sale of Goods and Supply of Services Act 1980

Under this Act goods must:
- **be of merchantable quality**: this means they should be fit for sale and not faulty
- **be fit for their purpose**: for example you should be able to walk in a pair of shoes without their falling apart
- **be as described**: for example if a food product is labelled 'gluten free' it should be so
- **correspond to (be the same as) samples**: for example you pick out a particular model of bathroom suite in a shop: the suite delivered to your house should be the same as the sample

Under this Act services, such as building and hairdressing, must:
- be carried out by someone with the necessary skills to do the job properly
- be carried out with care
- use good quality materials, for example a restaurant should use quality ingredients to make the food they serve

CONSUMER PROTECTION 16

Services must be carried out with care and skill

> If goods or services are faulty you are entitled to compensation; that is, repair, replacement or refund.

This Act does *not* cover you if:
- the fault was pointed out to you before the sale, for example goods reduced in price and marked as *seconds* or *imperfect*
- you misuse the goods
- there is no fault and you simply change your mind (some shops exchange goods anyway)

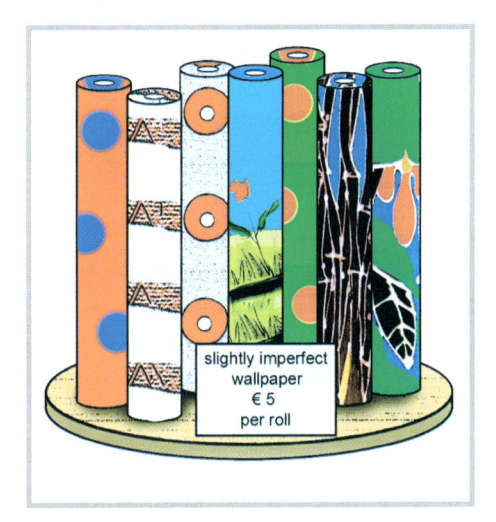

slightly imperfect wallpaper € 5 per roll

Activity 16.1 – Workbook p 142

A guarantee

A guarantee is a promise made to you by the manufacturers of the goods you buy. The guarantee, which must be written, usually states that the goods, if faulty, will be replaced or repaired free of charge. There is usually a time limit on the guarantee, for example one year. Guarantees are in addition to, not instead of, your rights under consumer law.

- No cash refunds.
- Credit notes only.
- No exchange on sale items.
- No receipt, no refund!

Notices like these are illegal if they refer to the refund or exchange of *faulty goods*. Remember if you simply change your mind about an item the shop does not have to refund your money or exchange the item.

The Consumer Information Act 1978

This Act states that information or claims made about goods and services must be true. The Act covers information in advertisements, on packaging, on shop notices and given by shop sales people. False claims about the prices of goods are also illegal under this Act (this includes present, previous and recommended retail prices).

ESSENTIALS FOR LIVING

Examples of claims made about goods and services

100% Wool

Guaranteed Irish

Genuine Leather

12 Hour photo

Costa del sol
To let
Apartment
3 mins from beach

We deliver within 30 minutes

Immaculate 3-year-old Nissan Primera only 15,000 miles.

Activity 16.2 – Workbook p. 144

Breaches of the Consumer Information Act should be notified to The Director of Consumer Affairs

Activity 16.3 – Workbook p. 145

Exam time – Workbook p. 146

Now test yourself at *www.my-etest.com*

Making a complaint

From time to time we all buy goods that turn out to be faulty. Knowing how to make a complaint effectively makes the experience better and less stressful for all involved.

When you first notice a fault:
- stop using the goods
- find your receipt
- know your rights
- make your complaint to the shop where you bought the goods as soon as possible

Remember! The shop where you bought the goods and not the manufacturer should deal with your complaint.

At the shop:
- ask for the customer services manager or, in a small shop, the manager
- calmly explain what is wrong
- state exactly what you want done, for example repair, refund or replacement
- if you get no satisfaction ask for the name of the managing director of the company, leave the shop and put the complaint in writing

Remember! No claim if:
- the fault was pointed out at time of purchase
- you abused the goods
- you simply changed your mind and there is no actual fault

Writing a letter of complaint

Write or type a neat letter to the managing director of the company. Photocopy the letter and keep the copy safe.

ESSENTIALS FOR LIVING

Sample letter of complaint

[Name and address of managing director of the company]

[Customer's address]

Mr Paul McManus
McManus Sports Ltd,
12 Market Street,
Waterford,
Co. Waterford

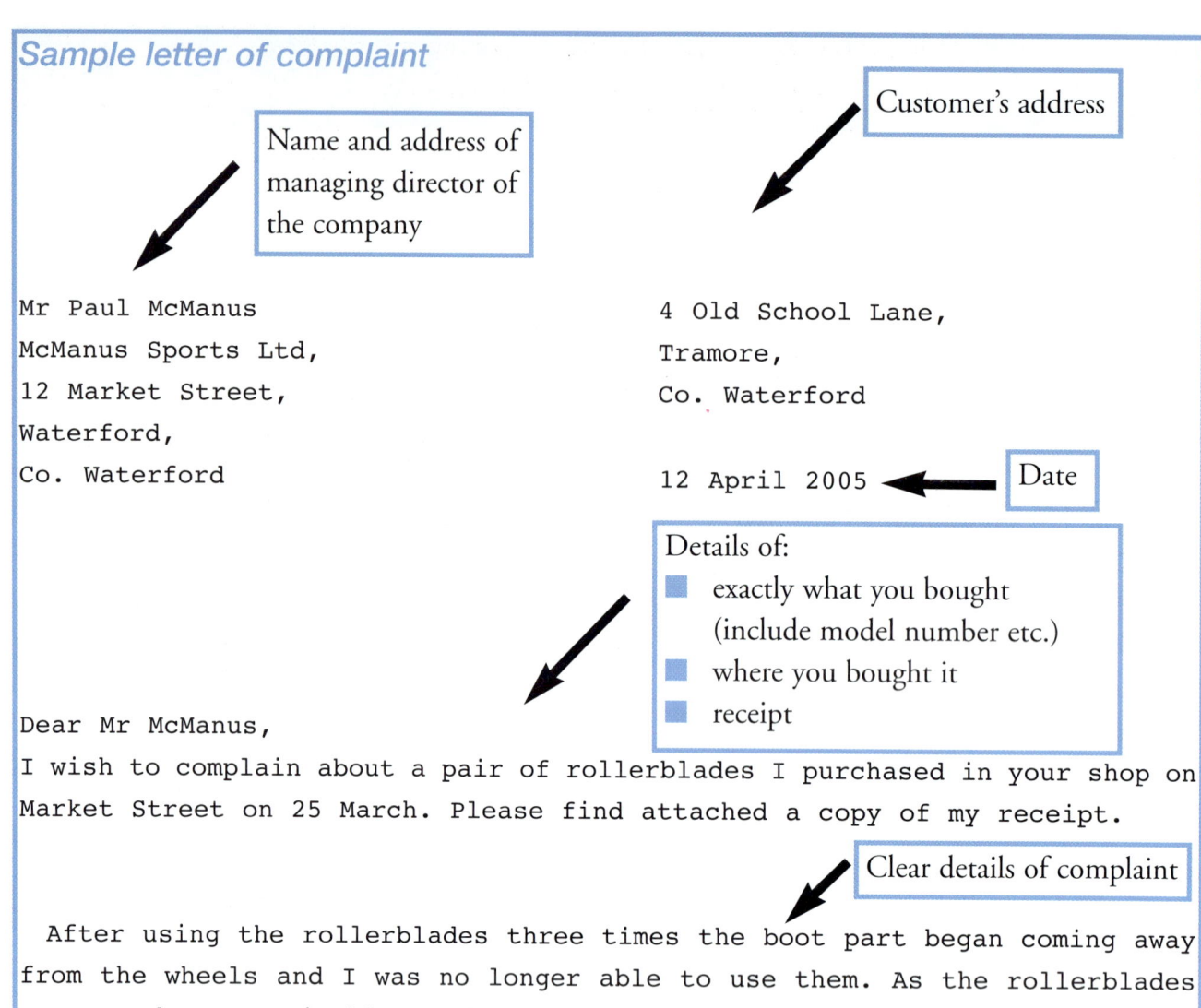

4 Old School Lane,
Tramore,
Co. Waterford

12 April 2005 — [Date]

[Details of:
- exactly what you bought (include model number etc.)
- where you bought it
- receipt]

Dear Mr McManus,
I wish to complain about a pair of rollerblades I purchased in your shop on Market Street on 25 March. Please find attached a copy of my receipt.

[Clear details of complaint]

After using the rollerblades three times the boot part began coming away from the wheels and I was no longer able to use them. As the rollerblades were used on a suitable surface the fault is most certainly with them. I returned the boots to your Market Street branch, but your shop assistant was unable to help me.

[Details of action you would like taken]

It is quite clear that there was a fault in the rollerblades when I bought them. I would like a refund for the total amount of €76.

Yours sincerely,

Peter O'Brien

Wait a few weeks for a reply; you could then follow up with a phone call. If you still get no satisfaction you may wish to go to a small claims court or to one of the following agencies for further advice.

Activity 17.1 – Workbook p 148

MAKING A COMPLAINT 17

Agencies that provide consumer information and advice	
Statutory (government-run) agencies	
Name of agency	**Function**
Office of the Director of Consumer Affairs and Fair Trade, Shelbourne House, Shelbourne Road, Donnybrook, Dublin 4	• enforces consumer laws • helps control the advertising industry • promotes consumer awareness • controls safety and labelling of food and textiles
The Ombudsman	The office of the Ombudsman deals with complaints against government-run organisations e.g. the health board, An Post. Complain to the organisation itself first and only to the Ombudsman as a last resort.
Non-statutory (voluntary) agencies	
Consumer's Association of Ireland	• advises consumers • provides consumer information e.g. teachers' packs, leaflets • has an information website • produces a monthly consumer magazine called *Consumer Choice*. This magazine gives unbiased information on goods and services currently on the market.
Trade organisations	Some groups of business people e.g. auctioneers (MIAVI), publicans (Vintners Association) come together and form a trade organisation. These organisations develop codes of good practice for their members and will deal with complaints.

Small claims court

To make a complaint in the small claims court, you must get and fill in an application form from the small claims registrar in your local district court. The small claims court deals with claims up to the value of €1269.70 (£1000).

Activity 17.2 – Workbook p 150

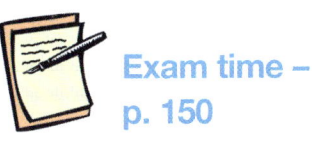

Exam time – Workbook p. 150

Now test yourself at *www.my-etest.com*

chapter 18

Quality

The term *quality* describes the standard of goods and services. Before goods leave a factory they are examined and tested to ensure that they reach a certain set standard. This is called *quality control*.

The quality and standard marks below are awarded to goods and services that are of a good quality or standard. These marks will be on the goods themselves or on their box or label.

This quality mark is awarded to Irish companies for quality goods and services and can be withdrawn if standards drop.	This symbol is awarded to high-quality goods made in Ireland.	This symbol is awarded to goods that are well-designed and made.
This Irish Standards mark is found mainly on electrical goods. It means that the goods have reached a high standard of safety and quality.	This mark is awarded by the EU to goods which have reached a high standard of safety. It is found on children's toys, clothes, buggies, electrical goods etc.	This kitemark is awarded to goods, e.g. electrical appliances, which have reached high standards of safety, performance etc.

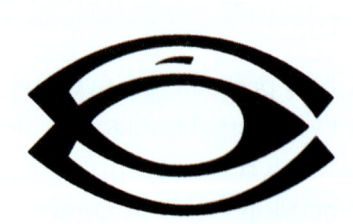

Activity 18.1 – Workbook p. 153

A good quality *service* will have:
- friendly staff who know what they are doing
- clean, well laid-out premises
- wheelchair access
- little queuing or waiting
- clean toilet facilities

Product labelling

Product labels should give you the following information:
- quality symbol
- warnings about the dangers of misuse
- name and address of manufacturer
- description of product, what it is made from, weight, colour, number of items etc.
- instructions for use, care and cleaning

Product safety

Product safety is concerned with two things:
- the safety of the user of the product
- that the product itself is not destroyed by being used, cared for or stored incorrectly

Product labels carry instructions and also warning and safety symbols. These try to ensure safe use of the product. The more common ones are shown in the table on page 166.

ESSENTIALS FOR LIVING

Instructions	Care labels	Food labels
Instructions tell you how to use the product safely. They may come as a separate booklet which should be kept in a safe place.	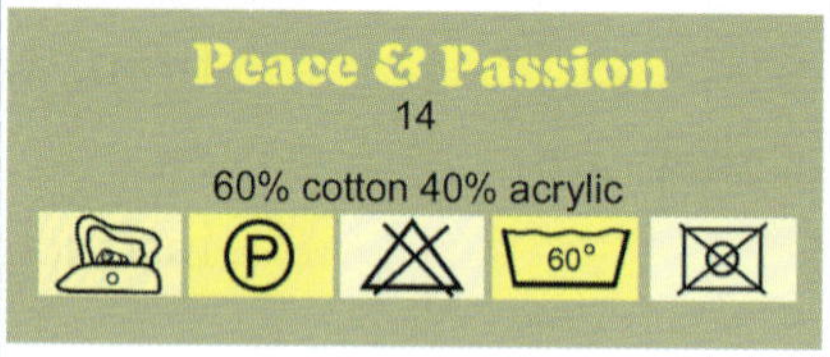 Care labels tell us how to wash, dry and iron the garment.	 Food labels Best before 30.01.04 Date stamping / expiry date

Safety symbols: show that products have been tested for safety

 BSI safety mark: awarded to safe electrical and gas appliances.	 Flame-resistant mark: found on furniture etc. Fabric carrying this symbol does not burn easily.	 European Standards mark: products carrying this symbol, e.g. children's toys and nightwear, have passed strict safety tests.	 Doubly insulated mark: found on small electrical appliances.

Warning symbols: found on potentially dangerous products

 Government warnings	 Harmful irritant: this product will damage the skin and eyes (e.g. bleach)	 Highly flammable: will burst into flames easily (e.g. aerosols)

Activity 18.2 – Workbook p. 153

166

QUALITY 18

Packaging

With the increase in self-service shopping (such as in supermarkets) and in the use of convenience foods, the amount of packaging has also increased.

Functions of packaging

Packaging:
- advertises the product
- makes it easier to store
- protects the product (e.g. food) from contamination
- preserves food (e.g. cans)
- provides essential information, e.g. instructions for use
- carries the barcode

Types of packaging

| Plastic | Paper |
| Glass | Metal |

Good packaging is:
- environmentally friendly
- strong
- light
- easily opened and resealed (if necessary)

Disadvantages of over-packaging

167

Too much packaging:
- is bad for the environment
- uses up natural resources (e.g. trees)
- causes litter and pollution (e.g. plastics)
- is expensive
- stops us seeing what we are getting for our money (e.g. ready meals)

(See Community services and the environment, Chapter 32)

Exam time – Workbook p. 154

Now test yourself at *www.my-etest.com*

chapter 19

Money management

Money management is planning our spending wisely so that we have enough for our needs and avoid debt. A *budget* is the actual plan for spending and saving. A good budget balances income (the money we have) with expenditure (what we spend).

Higher level

The money management process is continual and involves five stages. Budgeting is part of this process.

A TEENAGE BUDGET

1. Find out your exact income. Write it down.
2. List and price items you spend your money on. Also list savings.
3. Try to balance both figures. *Overspending* occurs if you borrow to buy things you cannot afford.

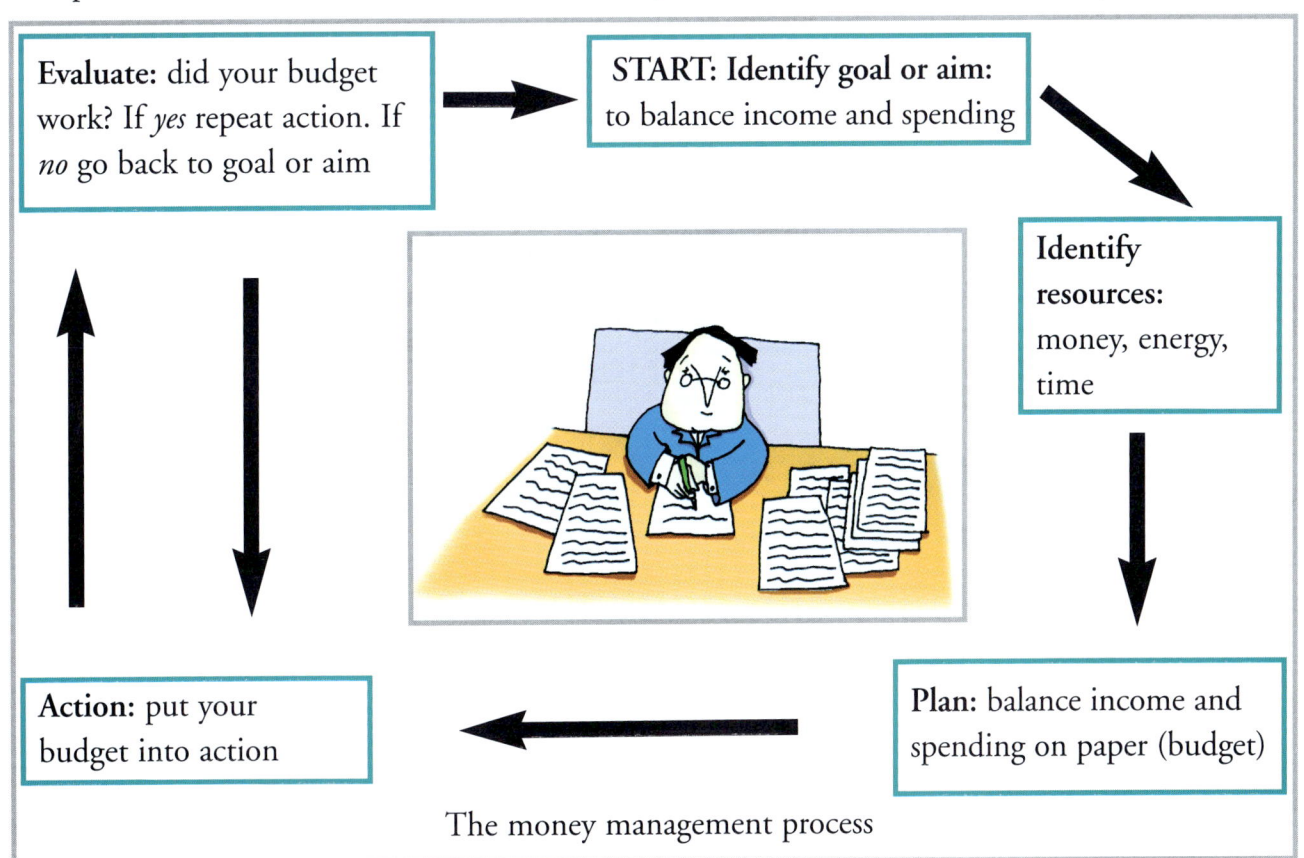

Evaluate: did your budget work? If *yes* repeat action. If *no* go back to goal or aim

START: Identify goal or aim: to balance income and spending

Identify resources: money, energy, time

Action: put your budget into action

Plan: balance income and spending on paper (budget)

The money management process

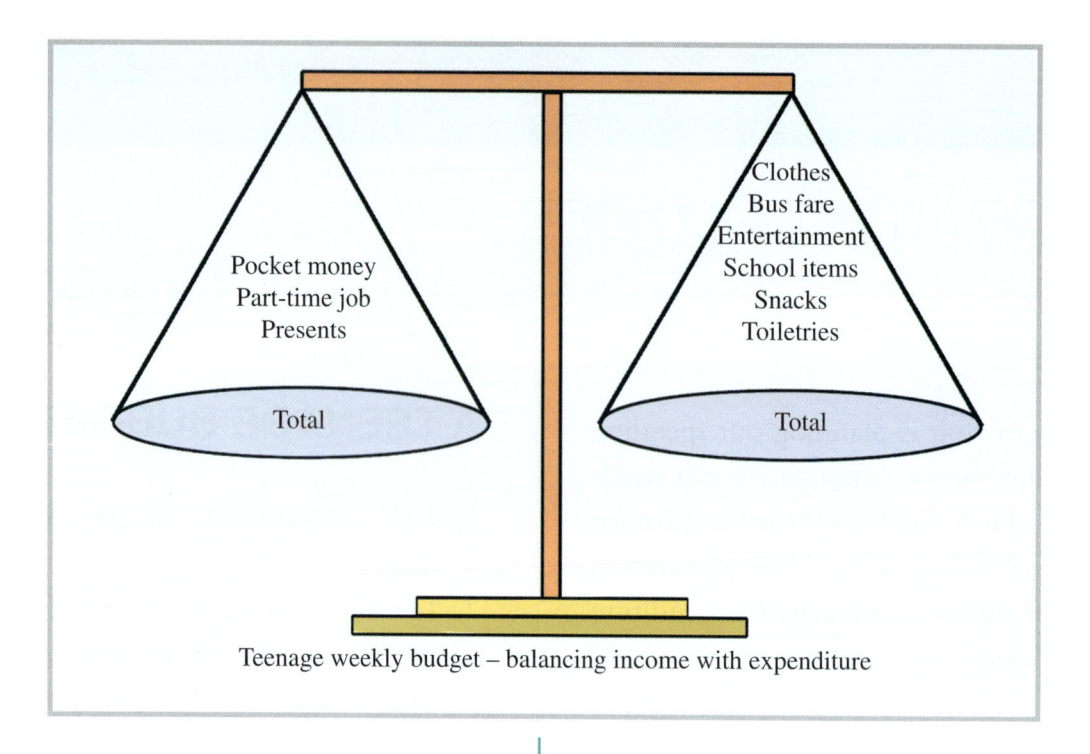

ESSENTIALS FOR LIVING

Teenage weekly budget – balancing income with expenditure

Teenage weekly budget: balancing income with expenditure

> Some costs, such as snacks, are weekly costs. For other costs, such as clothes, which you may not buy every week, an average weekly figure is worked out. For instance if you spend on average €40 per month on clothes, your average weekly spending on clothes is €10.

✏️ Activity 19.1 – Workbook p. 159

✏️ Activity 19.2 – Workbook p. 160

HOUSEHOLD BUDGET

The household budget must balance income with expenditure. Income may come from:
- earnings from employment
- state benefits, e.g. one-parent-family income, unemployment benefit

Household income

Income: Gross income *minus* deductions *equals* net income

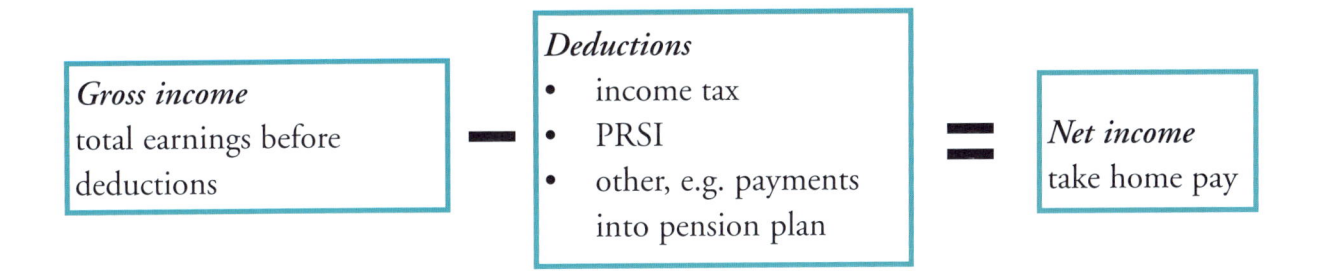

```
┌─────────────────────┐     ┌──────────────────────────┐          ┌──────────────────┐
│ Gross income        │     │ Deductions               │          │ Net income       │
│ total earnings      │  ─  │ • income tax             │     =    │ take home pay    │
│ before deductions   │     │ • PRSI                   │          │                  │
│                     │     │ • other, e.g. payments   │          │                  │
│                     │     │   into pension plan      │          │                  │
└─────────────────────┘     └──────────────────────────┘          └──────────────────┘
```

Income tax

Income tax is money that is either:

- deducted from your wages by your employer on your behalf and paid to the government (if you are a PAYE – pay as you earn – worker)
- paid directly to the government by you (if you are self-employed)

Income tax pays for essentials such as hospitals, schools, roads, gardaí etc.

Pay Related Social Insurance (PRSI)

A percentage of your gross income is taken in PRSI (most younger people are currently on the high rate which is four per cent [2003]). This money pays for benefits including some dental treatments, contributory pensions, benefits paid when you are ill or out of work.

Other income deductions

- pension payments: in permanent government jobs (e.g. teachers, nurses, gardaí) payments towards your pension (superannuation) are compulsory

Tax credits

From 6 April 2001 Ireland adopted a new tax credit system. Under this new tax system a person is given a certain number of tax credits per year depending on their circumstances.

Main tax-free credits (2002)

Single person   €1502
Married person   €3040
PAYE allowance   €660

Each week or month (depending on how you are paid) a person's *gross tax* is calculated. This gross tax is reduced by the *tax credit* due to the person for that week or month to get the net tax payable. Unused credits can be carried to the next week or month.

*Gross tax **less** tax credits **equals** net tax*

ESSENTIALS FOR LIVING

Household expenditure

The costs of basic needs such as food, clothing and shelter must be met first. Other necessities such as transport must then be met before luxury items are purchased.

Basic needs such as food must be provided for first in a household budget.
- food
- housing (rent or mortgage)
- household bills (e.g. electricity, gas)
- clothes and shoes
- transport (car or bus and train fares)
- crèche payments
- medical expenses (e.g. doctor, dentist, chemist)
- educational expenses (e.g. books)
- medical insurance (e.g. VHI, Bupa)
- house insurance
- luxuries (e.g. entertainment, holidays)
- savings

Activity 19.3 – Workbook p. 160

SAVING AND CREDIT BUYING

When you decide to buy something that you do not have enough money for, two options are open to you:
- save for the item
- buy it on credit

Saving

Saving for an item and then paying for it in full, in cash, is a good idea, for the following reasons:
- you earn interest rather than pay it
- the item costs less
- there is no risk of getting into debt

Where can you save?
- Credit Union
- bank or building society
- post office

Before deciding where to save find out:
- who offers the highest rates of interest?
- do they offer student incentives such as no bank charges?
- how easy is it to lodge and withdraw money?

In banks and building societies there are two basic types of account:

- deposit account (savings)
- current account

Money left in a deposit account will earn interest. A current account, on the other hand, is designed for everyday use. Money does not stay long in it and earns no interest. Bills can be paid by direct debit from a current account.

Opening a student deposit account

To open a student deposit (savings) account you must:

- fill out the bank or building society application form
- get a letter from your school saying you are a student there; the letter should also state your name, address and date of birth; *or* you can produce official identification such as your passport

Activity 19.4 – Workbook p. 161

Buying on credit

Credit buying means *buy now, pay later*. Credit encourages you to buy more than you can afford and has many disadvantages:

- interest is charged and the item ends up costing much more
- people run up debts

On the other hand, if you use credit wisely, you have the use of the item while paying for it. Some very large items could not be bought without credit, such as a house.

Where can you get credit?

1. Bank / building society loan
2. Credit Union loan
3. Bank overdraft: bank allows your account to go into the red; charges interest on money owed
4. Hire purchase: pay for goods in instalments; you don't own the item until it is paid for
5. Catalogues such as Family Album allow customers to pay for goods in instalments

Home filing

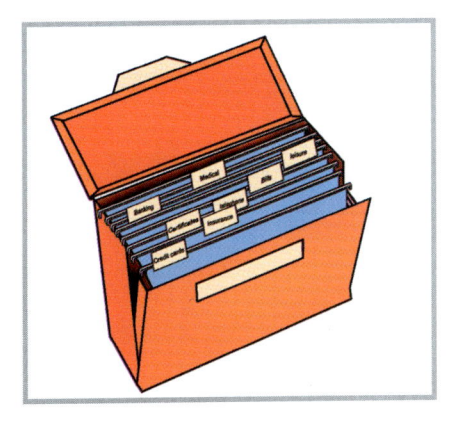

Keeping a household filing system is important for a number of reasons:

- bills are not mislaid and can therefore be paid on time

- time is not wasted searching for paper-work; it is all in one place
- guarantees and receipts are at hand should you have a problem with goods purchased

There are many simple yet effective home files on the market; generally they cost €20–40.

Exam time – Workbook p. 163

Now test yourself at *www.my-etest.com*

chapter 20

Shopping

How we shop has changed dramatically over the past twenty years. There has been:
- a decrease in the number of counter service shopping (e.g. corner shops)
- an increase in self-service shopping (e.g. supermarkets)
- a growth in the number of shopping centres
- increased late opening and Sunday shopping
- better facilities for shoppers (e.g. crèche, parking, cafés)

Self-service	Advantages	Disadvantages
	<ul><li>quick</li><li>products cheaper especially own brands e.g. Euroshopper and St. Bernard</li><li>good choice</li><li>high turnover means fresh goods</li><li>time is available to study and compare goods to make informed decisions</li></ul>	<ul><li>less personal</li><li>impulse buying more likely</li><li>sometimes crowded with long queues at checkouts</li><li>often on the outskirts of town so car needed</li></ul>
Counter service	**Advantages**	**Disadvantages**
	<ul><li>personal, friendly service: this is particularly important for those living alone e.g. the elderly</li><li>nearby: great for those without a car or if you need to pop out for something, e.g. milk</li><li>some pack and deliver groceries</li><li>some give credit</li></ul>	<ul><li>can be slow</li><li>less choice</li><li>slower turnover: goods may not be as fresh as supermarket products</li><li>more expensive</li></ul>

ESSENTIALS FOR LIVING

Types of shops		
Multiple chain store These are large self-service shops. Branches are all owned by the same company. Examples include Dunnes Stores, Penney's, Marks and Spencers, Tesco, NEXT, A-wear.	 **Voluntary chain stores** An individual shop owner can apply and pay to be part of a voluntary chain, e.g. Spar, Centra, Londus, Valuland.	 **Independent shops** Traditionally family-run shops, usually open till late and on Sundays. Limited choice of goods but personal service.
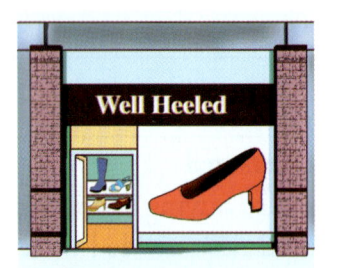 **Specialist shops** Usually sell only one type of product, e.g. shoes, cameras. A boutique specialises in expensive clothes.	**Department stores** Very large shops divided into different departments, e.g. household, ladies' fashions, gift wear, sportswear etc. Each department has its own sales staff with good product knowledge e.g. Clery's, Arnotts, Brown Thomas.	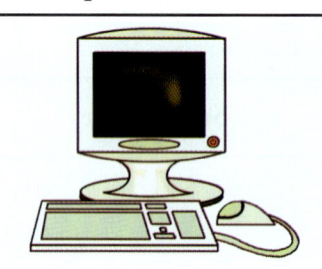 **Other** • shopping on line • discount stores: may also be voluntary chains, e.g. Valuland • mail order shopping e.g. Family Album • markets • auctions

Activity 20.1 – Workbook p. 167

Shopping guidelines

1. Make a list:
   - check what is in the house already
   - plan meals for the week
   - write up list: be as accurate as possible, e.g. write '4 tomatoes' rather than just 'tomatoes'
   - group similar items together, e.g. list all fruit and vegetables together as these are found together in the supermarket
2. Stick to the list: avoid impulse buying
3. Shop around for good value
4. Try not to shop too often: you end up spending more
5. Check expiry dates for freshness

SHOPPING TECHNOLOGY

Supermarkets today use various types of technology to make their service quicker and more efficient. For example:
- self-service weighing scales that print the product label
- computerised checkouts that scan *barcodes* and then print *itemised receipts*

Barcodes

A barcode is a series of lines and spaces that are read by a scanner attached to a computerised cash register.

Advantages and disadvantages of barcodes

- *itemised receipt* allows customers to see exactly what was bought
- quick stock control: once an item is scanned, a central stock control computer records that there is one less of that item on the shelves. When stocks of the item get to a certain level the computer re-orders the product automatically
- goods are not individually priced

Itemised receipt

Big Brother is watching!

Some large chains offer customer loyalty cards e.g. Superclub, Valuecard. The card is swiped every time you shop and you build up points towards various 'gifts'. The advantage for the shop is that they have your name and address together with a complete record of everything you have ever bought in the store. This information can then be used for marketing purposes. For example, a supermarket has a consignment of a particular brand of bolognese sauce that is not selling well: they see that you buy bolognese sauce weekly; they send you money-off coupons to encourage you to buy the brand they want rid of instead of your normal brand.

ESSENTIALS FOR LIVING

Methods used by supermarkets to encourage you to buy

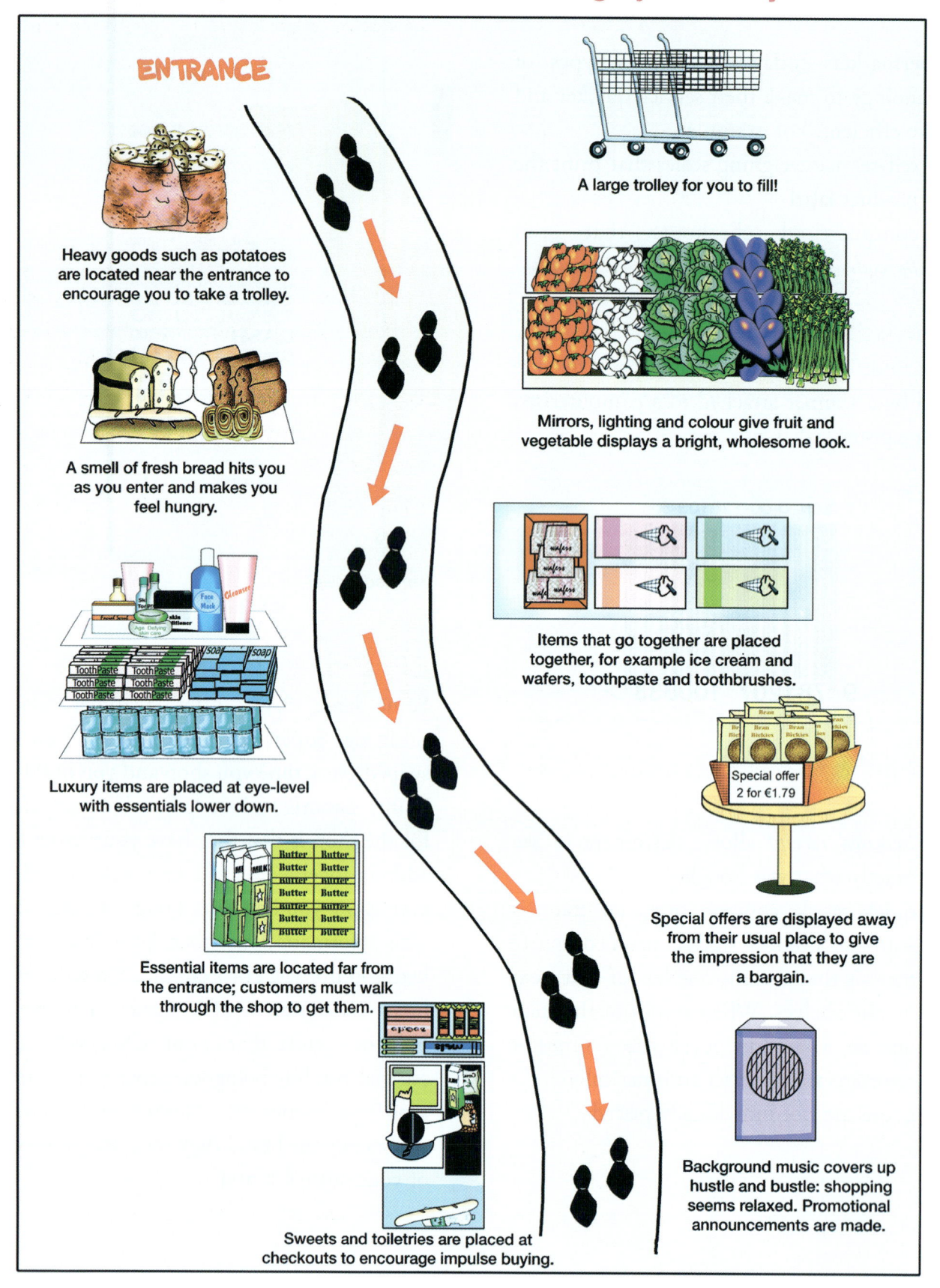

ENTRANCE

Heavy goods such as potatoes are located near the entrance to encourage you to take a trolley.

A smell of fresh bread hits you as you enter and makes you feel hungry.

Luxury items are placed at eye-level with essentials lower down.

Essential items are located far from the entrance; customers must walk through the shop to get them.

Sweets and toiletries are placed at checkouts to encourage impulse buying.

A large trolley for you to fill!

Mirrors, lighting and colour give fruit and vegetable displays a bright, wholesome look.

Items that go together are placed together, for example ice cream and wafers, toothpaste and toothbrushes.

Special offer 2 for €1.79

Special offers are displayed away from their usual place to give the impression that they are a bargain.

Background music covers up hustle and bustle: shopping seems relaxed. Promotional announcements are made.

Activity 20.2 – Workbook p. 168

Methods of payment

Cash is still the most common method of payment today. Some large supermarkets have ATM machines in-store for customers to withdraw cash. PASS and Banklink are examples of ATM cards.

Cheque and cheque card: Customers can write a cheque if they have a cheque guarantee card. The card guarantees the retailer that the cheque won't bounce. Cheque cards usually cover cheques up to €130.

Debit card (e.g. Laser): The card is swiped and the amount entered. The customer signs a receipt and this amount is taken from their bank account within one or two days.

Credit card (e.g. Visa, Access, MasterCard): The card is swiped and the amount entered. The consumer signs the receipt and the amount goes onto a bill. The bill can be paid in part or in full every month. No interest is charged if the bill is paid in full.

Activity 20.3 – Workbook p. 169

SHOPPING TERMS

Loss leader: These are goods sold off cheaply by the retailer. It is hoped that these will lead you into the shop where you will buy more.

Bulk buying: You buy a product in large quantities: cheaper so long as you use it all.

Unit pricing: Goods are priced per unit (e.g. per kilogram). Meat, fruit and vegetables are often priced in this way.

Own brands: Own brands are plainly packaged products (e.g. St. Bernard, Euroshopper). They are often good quality and much cheaper than branded products (e.g. Lyons Tea).

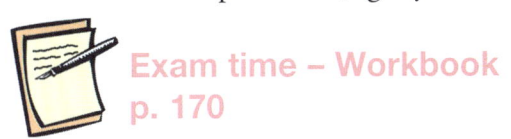

Exam time – Workbook p. 170

Now test yourself at *www.my-etest.com*

chapter 21

Advertising

Advertising is a feature of modern life. We are constantly exposed to it.
- TV, radio
- newspapers, magazines
- cinema, video
- billboards, bus shelters
- buses, trains
- sports grounds, sports sponsorship
- direct mail
- carrier bags, logos on clothing etc.

An effective advertisement must:
- catch your attention
- make you want the product
- persuade you to go out and buy the product

Advertising techniques

There are **eight** basic advertising techniques used by companies to encourage you to buy their products.

1. Humour catches your attention

2. Glamour and romance: 'If you buy this product you too can look like this/find romance'

180

ADVERTISING 21

3. Traditional wholesome images: 'This product is healthy'

4. Mr and Mrs Average: 'This person is just like you and finds this product wonderful'

5. Happy families: 'This product will make your family happier'

6. Famous personalities: 'If this rich and famous person uses this product it must be good!'

ESSENTIALS FOR LIVING

7. Slogans: 'The best-built cars in the world'

8. Statistics: 'Eight out of ten cats prefer it'

PETS

ABANDONED ANIMALS need kind homes for labrador (f) 1 yr. spayed; puppy, 12 weeks (m) house-trained, lovely nature; border Collie (m) 2 yrs; kittens and young adult cats. Tel: 1234567

LOVING HOMES wanted for whippet (m) 4 yrs; black & white Cocker Spaniel (f) 6 yrs; Bloodhound (m) 5 yrs; all dogs very loving with excellent temperaments. Foster homes and fund raisers urgently needed. Tel: 3456789

PET RESCUE urgently need kind homes for Dalmation (m) 7 yrs, owner deceased; pair of gentle Collies (f) 2 yrs; German Shepherd (m), 3 yrs, house-trained; Sable Terrier (f) 1 yr, owner deceased; King Charles Spaniel (f) 10 months, very gentle; Long-haired Lurcher (like Wolfhound) (f) 2-3 yrs; black Setter type (f) 6 mths; tri-coloured Terrier (f) 2 yrs; affectionate and needy cats and kittens - all colours. Tel: 2345678

Classifieds are ads placed in newspapers and magazines. They are used to sell goods such as cars and furniture and to let and sell property etc.

Activity 21.1 – Workbook p. 174

Advantages and disadvantages of advertising

Advantages	Disadvantages
• provides information • creates jobs in advertising • creates jobs in factories and shops because of increased sales • keeps the cost of TV, newspapers, magazines down	• heavily advertised goods are often expensive e.g. Levi jeans • encourages people to want only famous brand names • encourages overspending • sometimes stereotypical images are used e.g. women really concerned that their whites are white • spoils the landscape e.g. billboards • interrupts TV viewing

Advertising control

Because everyone in our society is exposed to advertising it must be controlled. Advertising is controlled by (i) the law, (ii) the industry itself.

The law

Consumer Information Act, 1978: Advertisements must be truthful

Employment Equality Act: Protects against discrimination such as 'single female required for busy restaurant'

Voluntary control

Advertisers are encouraged by the Advertising Standards Authority to produce ads that are legal, decent, honest and truthful.

Marketing

Nowadays marketing plays a big part in industry.

Marketing is finding out customer's wants & needs and then satisfying them...

Market research means gathering information about consumers' likes, dislikes, wants and needs. Information is gathered by questionnaire, phone interviews and through the use of club cards etc.

ESSENTIALS FOR LIVING

Case study:

Through market research Pepsi Cola found that both males and females were concerned with the high levels of sugar in soft drinks. Males were, however, unwilling to be seen as 'wimps' by drinking diet drinks. Pepsi overcame this by creating a diet drink with a macho image – Pepsi Max.

Activity 21.2 – Workbook p. 174

Exam time – Workbook p. 176

Revision crossword – Workbook p. 177

Now test yourself at *www.my-etest.com*

Unit 3

Social Studies

chapter 22

The family and adolescence

THE FAMILY

Types of family

Nuclear family: father, mother, child or children living in the same house

Extended family: includes grandparents, uncles, aunts and cousins all living together or near each other

Single-parent family: family headed by a parent who may be unmarried, separated, divorced or widowed

Sibling: brother or sister

Functions of the family

A *functional* family will provide us with all our needs. A *dysfunctional* family will only provide some of them.

Needs

Physical needs: food, clothing, shelter
Emotional needs: love, security, support

'Mammy and Daddy do be fighting. I don't be afraid cos I'm big, I'm five and a half.'

What need is not being provided for the child who said this? Do you think he or she is afraid?

Changes to the family

Single parents: In the past in Ireland the only type of single parent family seen as acceptable was one where a parent had died. Today, as a result of divorce, increased marital separation and an acceptance of single unmarried parents, many Irish households are headed by a single parent.

Single-parent families are very common in modern Ireland

Cohabiting couples: Many couples, for various reasons, decide not to marry but live together and rear their children.

THE FAMILY AND ADOLESCENCE

Isolation: Nowadays, many couples move away from their extended family to find work etc. Child-rearing can be more difficult as there are no family members nearby to help out.

Roles and relationships

Each person within a family has a role. Roles include those of the child, adolescent, parent or grandparent. The role we have tells us how we are expected to behave within the family.

| Examples of roles and expected behaviours ||
Role	Expected behaviour
Children	• to play, to go to bed early
Adolescents	• to do their share of the housework, to study for exams, to tell parents where they are going
Parents	• to provide for their children's needs

Relationships between family members will generally be good if each person fulfils their role. There is often conflict between adolescent family members (teenagers) and their parents because of role confusion.

Activity 22.1 – Workbook p. 181

GENDER ROLES

Gender means being either male or female. In the past males and females had very different *gender roles*, meaning they were expected to behave very differently from each other. For example, many men were expected to go out to work and do very little housework or child-rearing. Women were expected not to work outside the home and to do all the housework and child-rearing.

Stereotyping

Stereotyping means having an oversimplified, fixed, and often wrong idea about a group of people, for example 'All Americans are loud.'

Gender stereotyping means having a fixed and often wrong idea about men's and women's personalities and life expectations.

> **Examples of gender stereotyping**
> Personalities: women are emotional; men are not.
> Life expectations: women just want to find a husband and have children; men want a career.

Gender-role stereotyping means having a fixed and often wrong idea about how men and women should behave in society, for example 'Women should do the housework, men should do the garden.'

Activity 22.2 – Workbook p. 181

Gender equity is the opposite of gender-role stereotyping. In a world where there is gender equity both men and women can be who and what they want to be without being judged or discriminated against. Are men and women treated equally in Irish Society?

Many women around the world do not have the same equality and freedom of opportunity that the majority of us enjoy. In some Muslim countries, such as Afghanistan, Saudi Arabia and Pakistan, women must dress in a shroud-like burqa veil. The veil allows only the hands and feet to be visible. From 1996 to 2002, while the Taliban were in power in Afghanistan, women were not allowed to receive education, drive or work. Small things that Western women take for granted were banned, such as makeup, nail polish, jewellery, short hair and high heels.

An Afghan woman in a burqa

Exam time – Workbook p. 183 – Family roles and relationships

ADOLESCENCE

Growth and development

Our growth and development as human beings are influenced by two things:
- heredity: what has been passed on by our parents
- environment: how we are brought up; our surroundings

During our lives we go through four stages of development.

Childhood

Adolescence

THE FAMILY AND ADOLESCENCE 22

Adulthood

Old age

As human beings there are also four different *areas* of development.

Physical development

Mental development

Emotional development

Social and moral development

Type of development	What does it mean?
Physical	• growth of the body, e.g. getting taller • increased co-ordination and ability to do things, e.g. ride a bike
Mental	• development of memory and understanding • development of speech, reading, writing
Emotional	• learning to handle your emotions, e.g. anger, fear • development of self-esteem (what you feel about yourself)
Social and moral	• learning behaviours acceptable in society, e.g. sharing • developing friendships and relationships with others • knowing right from wrong

For development during childhood see childcare option page 305.

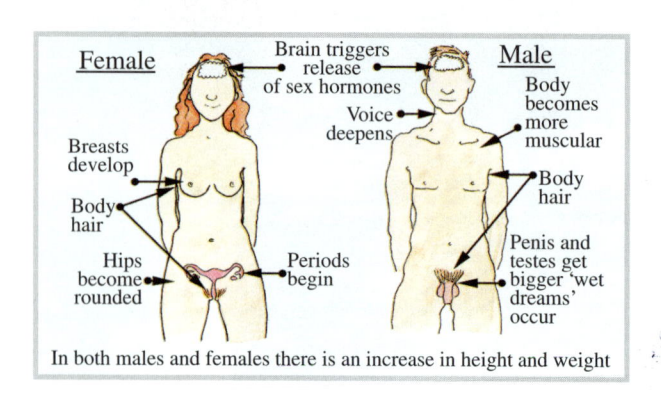

Female: Brain triggers release of sex hormones; Voice deepens; Breasts develop; Body hair; Hips become rounded; Periods begin.

Male: Body becomes more muscular; Body hair; Penis and testes get bigger 'wet dreams' occur.

In both males and females there is an increase in height and weight

Development during adolescence

Adolescence (age 12–18), like early childhood, is a time of rapid development.

Physical development

■ co-ordination improves: ability to dance, to play sports, to complete skilled tasks increases
■ physical and sexual development takes place during puberty, caused by sex hormones

Mental development

During adolescence most young people become capable of more complex thought. Younger children tend to think more in the here and now, about what they can now see, hear, smell or touch. They don't think about the past and the future as much, for example a child will rarely talk about his or her school day. Most adolescents, on the other hand, become able to:

■ plan, for example 'What will I do this weekend?'
■ handle long-term tasks, for example

THE FAMILY AND ADOLESCENCE **22**

Junior Certificate Home Economics options project
- think about 'what if' situations, for example 'If there is an accident at Sellafield, what *would* happen here in Ireland?'
- think about or do more than one thing at a time

Emotional development

During adolescence:
- sexual feelings develop
- adolescents are often very concerned about how others see them: an imaginary audience. The advertising industry uses these insecurities to sell their goods
- conflict increases

Conflict is a normal part of development for adolescents. Conflict is usually about housework, money, appearance, substance use or abuse (such as drinking), school work, curfew, boy/girl friends, and sexual behaviour. Adolescents usually want more independence than their parents will allow. This is frequently the cause of conflict.

Dealing with conflict

If you are fighting with a parent or someone else:
- *do not* get aggressive or blame them
- *do not* ignore the problem: it will not go away
- *do* calmly discuss the issue and try to work out a solution

Activity 22.3 – Workbook p. 182

Social and moral development

During adolescence the *peer group* becomes very important. Your peer group are people your own age who share similar interests. Adolescents usually spend less and less time with their family and more with their peer group. *Peer pressure* is when your peer group puts pressure on you to do something, for example to smoke. Peer pressure is not always bad. Can you think of an example of positive peer pressure in your life?

Moral development

Moral development occurs in two stages. Some people never reach stage two.

Stage 1: ruled by praise and punishment, for example 'I won't steal from my mother's purse because if she notices she will kill me.'

Stage 2: ruled by your own sense of right and wrong, for example 'I won't steal from my mother's purse because I feel it is wrong to steal.'

ESSENTIALS FOR LIVING

Norms

Higher level

Our sense of right and wrong is often influenced by the norms of the society we live in. Norms are acceptable ways of behaving, for example putting rubbish in the bin, not damaging other people's property etc.

Activity 22.4 – Workbook p. 182

SEX EDUCATION

The female reproductive system

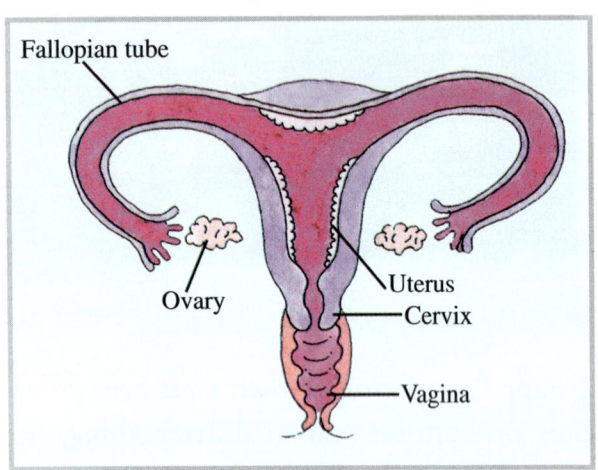

Diagram labelled: Fallopian tube, Ovary, Uterus, Cervix, Vagina

1. From puberty onwards an egg or *ovum* is released from one of the female's ovaries every month. This is called *ovulation*.
2. The egg goes into the *fallopian tube* nearby and waits to be fertilised by the male sperm.
3. If the egg is not *fertilised* it travels down into the womb and out of the body. The girl or woman then has her monthly period.
4. If the woman has sex, sperm comes up to the fallopian tube and joins with the egg: this is called *fertilisation* or *conception*.
5. The egg makes its way to the *uterus* (womb) and sticks itself to the womb wall. The woman is then pregnant.
6. The female hormones *progesterone* and *oestrogen* cause:
   - ovulation
   - changes during puberty, e.g. breasts, growth of pubic hair
   - changes during pregnancy

THE FAMILY AND ADOLESCENCE 22

Menstruation (periods)

Sanitary towels and tampons		Hygiene during menstruation
Sanitary towel with wings self-stick wings hold the towel in place and prevent leaking	*Pantyliner* for light bleeding or general freshness. Self-stick panel	• change pad or tampon every 3–4 hours • wash pubic area every day • wash your hands before and after changing a pad or tampon • wear tight-fitting underwear with pads to prevent leaking • buy dark-coloured underwear to wear during periods
Tampon small; must be inserted with fingers – takes practice to insert far enough	*Tampon with applicator* applicator pushes tampon far into vagina	
Using sanitary towels worn inside tight-fitting underwear		*Using tampons* worn inside the vagina; a string hangs outside the vagina to remove the tampon

193

ESSENTIALS FOR LIVING

The male reproductive system

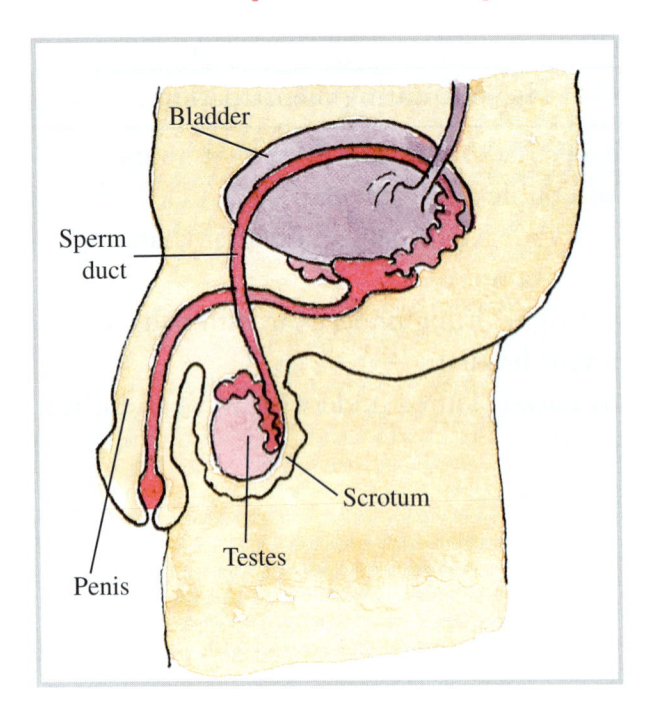

Diagram labels: Bladder, Sperm duct, Penis, Testes, Scrotum

1. From puberty onwards the male *testes* produce *sperm*. The testes lie outside the body in a sac called the *scrotum*.
2. When a male reaches orgasm during sexual intercourse, sperm travels quickly up the sperm duct and out through the penis.
3. Many thousands of sperm are released near the woman's cervix. They begin swimming frantically towards the fallopian tubes in an effort to fertilise an egg which may or may not be there. In the end only one sperm fertilises the egg.
4. The male hormone *testosterone* is made in the testes and causes:
   - sperm production
   - other changes during puberty, e.g. facial hair

BEING RESPONSIBLE

In Ireland in 2001 sixty-six young girls under sixteen had babies (CSO 2002). The legal age of consent (to have sex) in Ireland is now seventeen. Studies show that many teenagers (especially teenage girls) have sex because of various pressures rather than really wanting to. Most people believe that sex is best kept for close, long-term relationships and that casual sex can be damaging for the individual.

Casual sex can lead to:
- unexpected pregnancy
- feelings of being used
- sexually transmitted diseases, e.g. genital warts, herpes, AIDS

Alcohol or other drugs are often involved when people make risky decisions about sex.

Activity 22.5 – Workbook p. 185

Activity 22.6 – Workbook p. 187

THE FAMILY AND ADOLESCENCE   **22**

Pregnancy

Start: Fertilisation takes place in the fallopian tubes → After a few days the egg travels to the womb where it implants or sticks to the womb wall

↓

Sometimes the embryo does not stay implanted and an early miscarriage occurs: some women don't even know they are pregnant and think it's just a heavy period

↓

The *placenta* forms where the embryo is implanted or stuck. The placenta is the baby's lifeline. The baby is joined to the placenta by the *umbilical cord*

↓

From eight weeks the embryo is called a *foetus*. The placenta supplies the foetus with food and oxygen and takes away waste

↓

Unfortunately some harmful substances, such as alcohol, nicotine, heroin, cocaine and cannabis, and some viruses such as rubella and HIV can pass to the baby as well

←

The baby usually stays in the womb for *forty weeks after the first day of the woman's last period*. This is the *due date*

←

From about thirty-two weeks the baby's head is facing downwards. When the baby is about to come the waters usually break and contractions start: this is called *labour*. Some women have an *epidural*, which is injected into the spine for pain relief. Contractions get stronger and more frequent

↑

The head is delivered first. Sometimes a cut is made in the vagina to make the opening bigger and to stop tearing: this is called an *episiotomy*. Once the head is out the rest of the body and the *afterbirth* come

↑

The baby's umbilical cord is cut and clamped and the baby checked

↑

Approximately ten per cent of babies are born by Caesarean section. Here the woman's stomach and womb are cut open and the baby is removed. Usual reasons for Caesarean sections are large baby / small pelvis / breech baby (feet first)

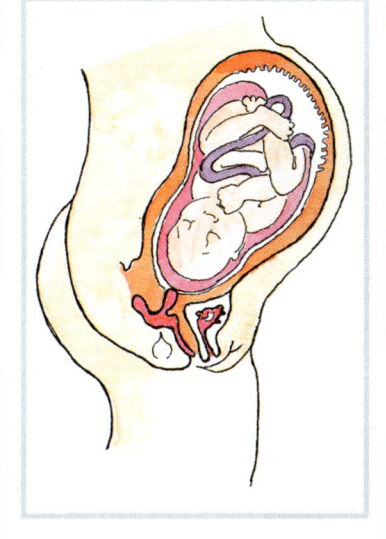

Exam time – Workbook p. 188 – Sex education

Now test yourself at *www.my-etest.com*

chapter 23

The human body

- the teeth
- the circulatory system
- the respiratory system

THE TEETH

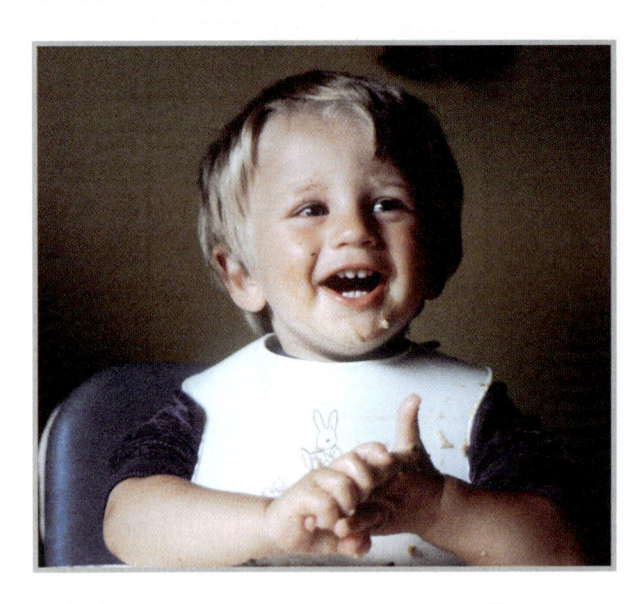

A baby's twenty milk teeth generally appear by age two and a half and fall out from age seven.

An adult has thirty-two permanent teeth. The four wisdom teeth are the last to appear, in the mid- to late teens.

Types of adult teeth

There are four different types of teeth, each with a different function.

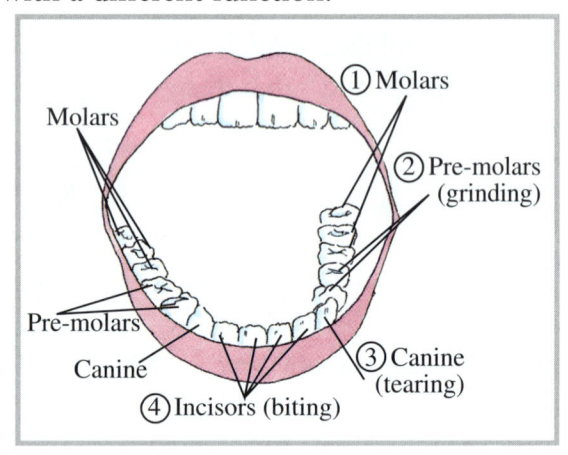

Types of teeth

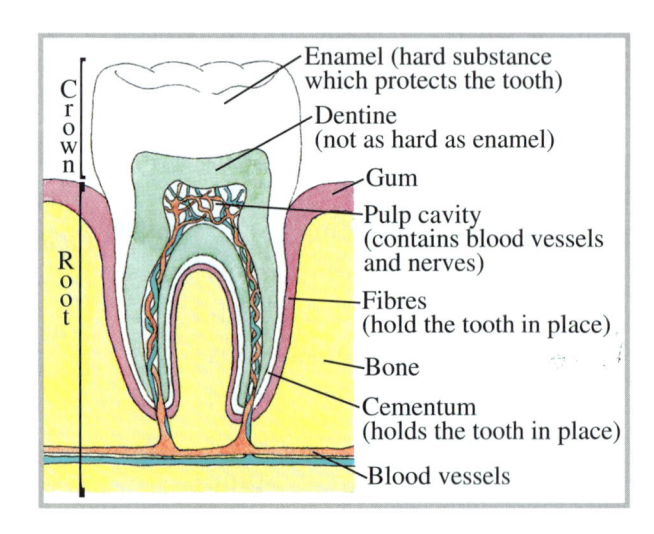

Structure of a tooth

Tooth decay and gum disease

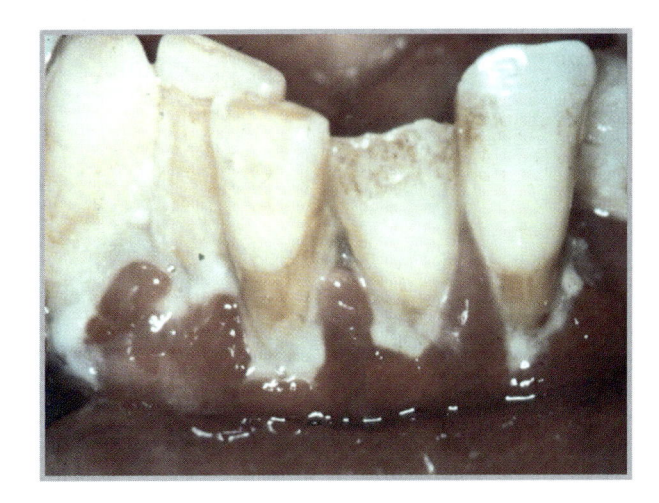

Having healthy teeth is very important. Tooth decay causes toothache, bad breath and gum disease and can make talking and smiling difficult or embarrassing.

Plaque
Plaque is the yellowish coating that forms on teeth between brushings. Plaque contains food and bacteria. If teeth are not brushed the bacteria begin digesting the food and *acid* is produced. This acid burns into the tooth and causes it to decay or rot. A toothache is not felt until the acid has burned through to the pulp cavity. Why?

Gum disease
Gum disease occurs when acid attacks the gums leaving them infected, sore and swollen.

Fluoride
Fluoride is a substance often added to water supplies and toothpaste. Fluoride strengthens teeth and helps prevent decay.

Caring for the teeth and gums
There are four basic rules for caring for the teeth and gums.

1. Eat calcium-rich foods and raw fruit and vegetables.

2. Avoid sugary food and drinks.

3. Brush after meals with a soft toothbrush. Replace brush regularly.

ESSENTIALS FOR LIVING

4. Have a dental check-up every six months. Never use your teeth to open bottles etc. This damages the enamel.

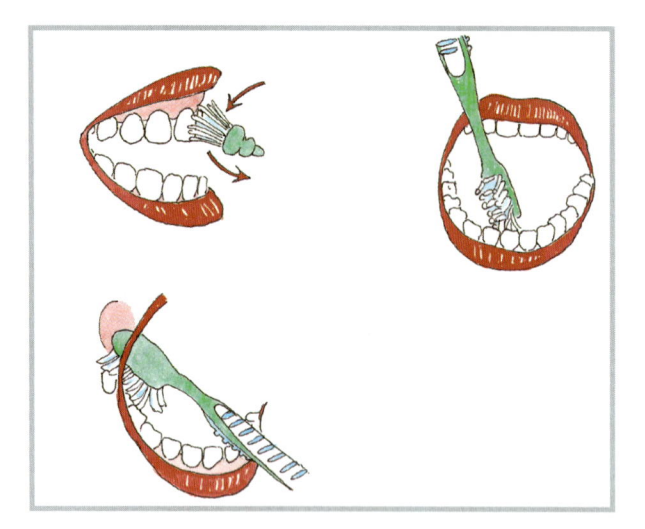

Dental products

Use dental floss to clean between the teeth safely. Use before brushing. Never use metal objects to do this: it damages the enamel.

Mouthwash helps kill the acid-producing bacteria in plaque. Use after brushing.

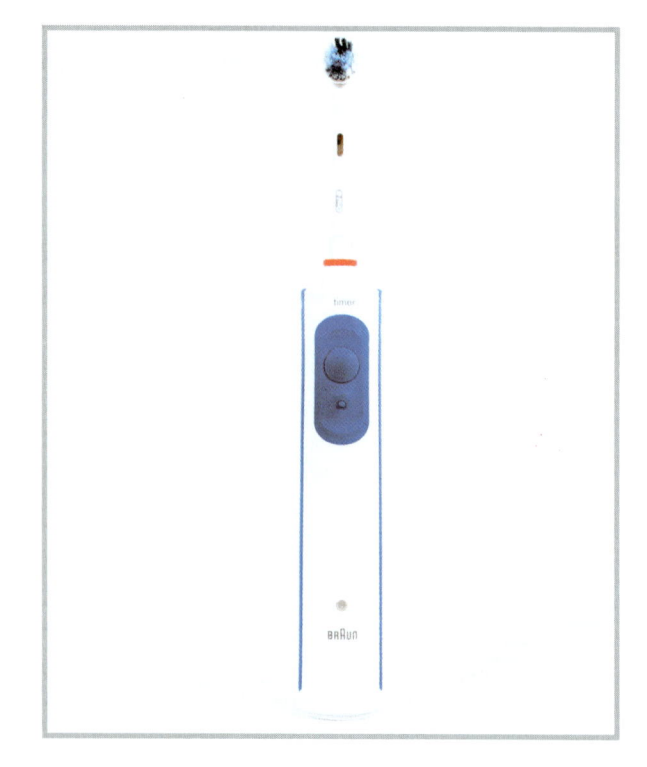

The good circular action of an electric toothbrush does not damage gums. New heads are expensive.

198

Activity 23.1 – Workbook p. 191

Exam time – Workbook p. 191 – The teeth

CIRCULATORY SYSTEM

Higher level

The circulatory system consists of the heart, the blood vessels and the blood. The job of the circulatory system is to carry blood full of oxygen and nutrients around the body and then to collect waste products, such as carbon dioxide.

The heart

The heart is a strong muscular organ that pumps blood either to the lungs (to get oxygen) or around the body. It lies between the lungs in the middle of the chest and is about the size of the owner's fist. The heart has four chambers with a wall called the septum dividing left and right sides.

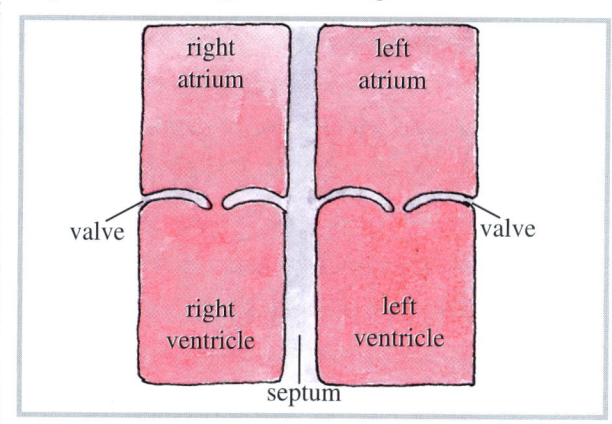

Four chambers of the heart

Blood-flow through the heart

1. Impure blood (full of carbon dioxide) comes into the heart from the rest of the body through the vena cava into the right atrium. The right atrium contracts (squeezes in) pushing the blood through valves into the right ventricle.
2. The blood is then pushed out of the heart through a valve into the pulmonary artery and to the lungs.
3. In the lungs carbon dioxide is exchanged for oxygen.
4. The blood then returns through the pulmonary veins to the left atrium.
5. The left atrium contracts pushing blood through a valve into the left ventricle.
6. The left ventricle then contracts pushing blood through a valve into the aorta.
7. The aorta brings the oxygen-rich blood around the body.
8. The body uses this oxygen and produces carbon dioxide. This carbon dioxide is brought back to the vena cava and the cycle begins again.

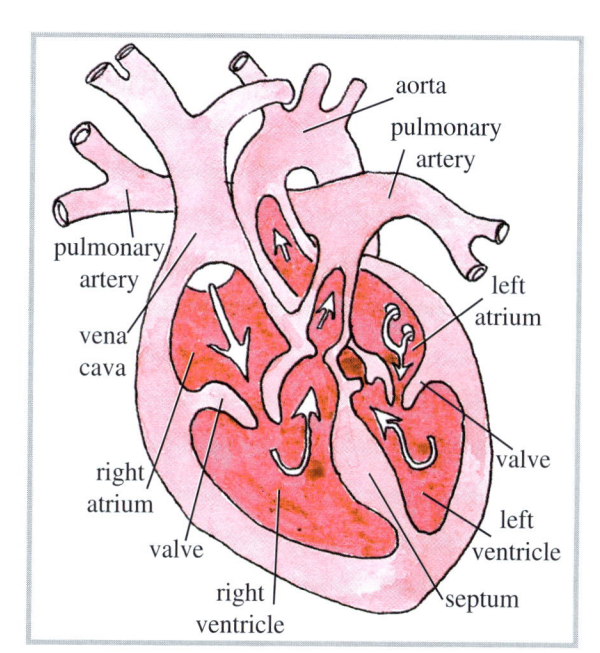

Activity 23.2 – Workbook p. 193

ESSENTIALS FOR LIVING

The heartbeat is the sound of the valves in the heart opening and closing. Each time the heart beats blood is forced into the arteries; this can be felt as a pulse. The average pulse is seventy-two beats per minute and can be felt anywhere an artery is close to the skin's surface, for example in the wrist or under the ear below the jawbone.

Activity 23.3 – Workbook p. 193

Blood vessels

Blood vessels are tubes which carry blood around the body. There are three types:

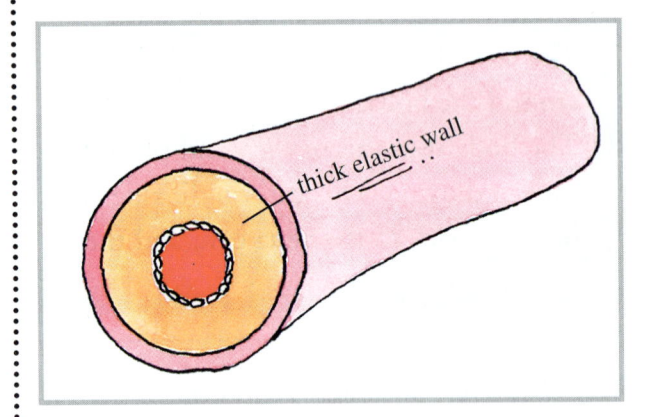

1. Artery

Arteries carry blood away from the heart under pressure. They have thick elastic walls.

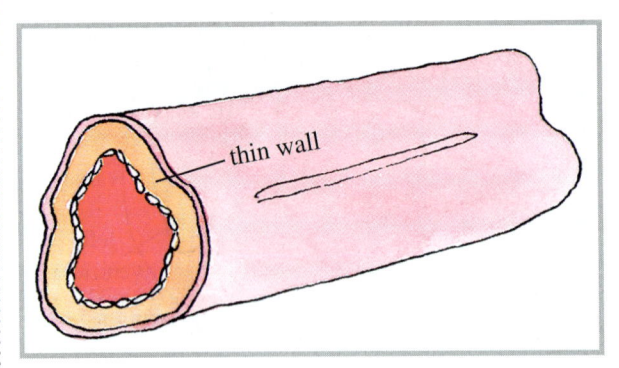

2. Vein

Veins carry blood to the heart. The blood is not under much pressure so valves are needed to stop it flowing backwards. They have thin walls.

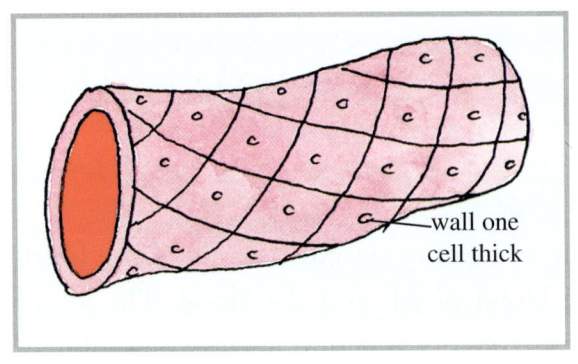

3. Capillary

Capillaries join arteries and veins. Their walls are only one cell thick. This allows oxygen and nutrients to pass through into the body cells.

The blood

Composition

Plasma is yellowish liquid in which the blood cells float.

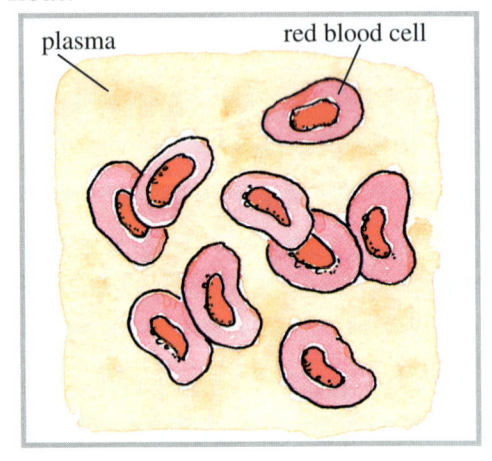

Red blood cells: these disc-shaped cells contain haemoglobin which carries oxygen around the body. Iron is needed to make haemoglobin.

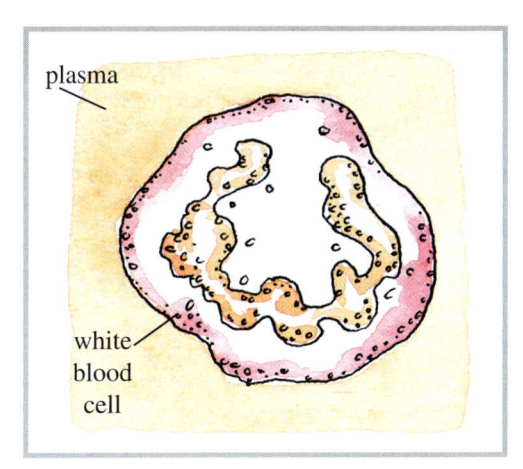

THE HUMAN BODY

RESPIRATORY SYSTEM

The respiratory system takes oxygen from the air when we breathe in and brings it to the cells of the body to make energy. This is called oxidation. Carbon dioxide is produced as waste and is removed by the respiratory system when we breathe out. The respiratory tract (below) is part of this system.

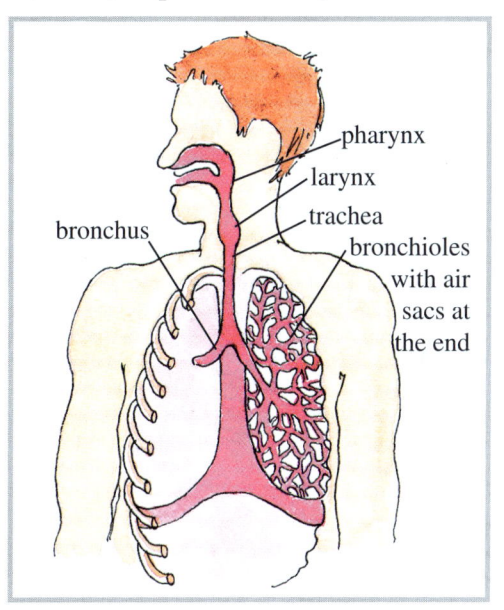

White blood cells: these fight infection. They surround the germs causing the infection and destroy them. Puss is the white blood cells and germs fighting each other.

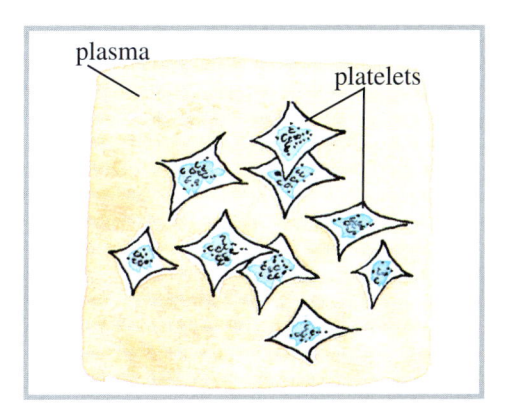

Platelets: small cells that clot blood when we get cut, to form a scab.

Functions of the blood
- transports oxygen, carbon dioxide, nutrients, hormones, enzymes and waste around the body
- fights infection (white blood cells)

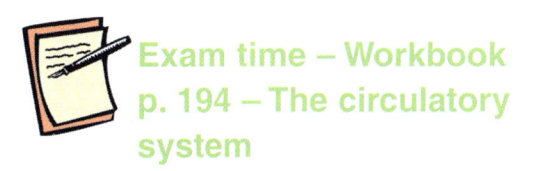

Exam time – Workbook p. 194 – The circulatory system

1. Air passes through the mouth and nose where it is warmed and some dust is filtered by the hairs in the nose.
2. Air then passes by the *pharynx* and *larynx* (voice box) and *epiglottis*. The epiglottis is a flap of skin that covers the windpipe when we swallow to stop food going down the windpipe.
3. Air then passes into the *trachea* or windpipe. The trachea divides into branches called *bronchi*. Each bronchus divides again and again to form tiny *bronchioles*.
4. At the end of each tiny bronchiole is an *air sac* or *alveolus* (plural: *alveoli*). Here oxygen and carbon dioxide are exchanged.

Exchange of gases

The exchange of gases happens in the air sacs of the lungs. The air sacs are surrounded by capillaries. Carbon dioxide is removed from the blood in the capillaries and oxygen is added in its place. This oxygen-rich blood then returns to the heart and is pumped around the body.

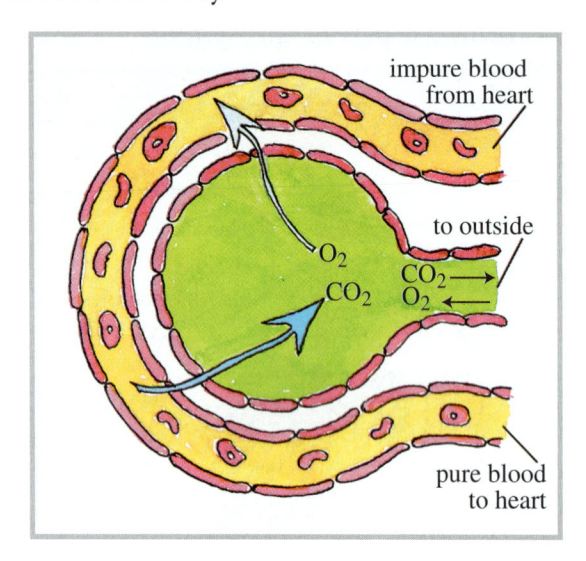

Activity 23.4 – Workbook p. 194

Functions of the lungs
(Ordinary and Higher levels)

- take in oxygen
- remove carbon dioxide and water vapour

(Diseases of the lungs: see Smoking page 208)

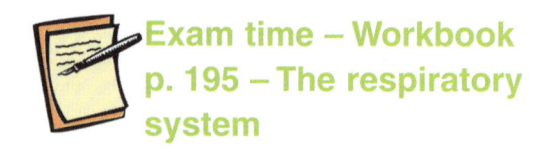

Exam time – Workbook p. 195 – The respiratory system

Revision crossword Workbook p. 196

Now test yourself at *www.my-etest.com*

chapter 24

Health education

- good health
- health hazards
- the skin and personal hygiene

HEALTH

The World Health Organisation (WHO) describes health as *a complete state of physical, mental and social well-being and not simply the absence of disease or infirmity*. With this definition in mind we can list seven general guidelines to good health.

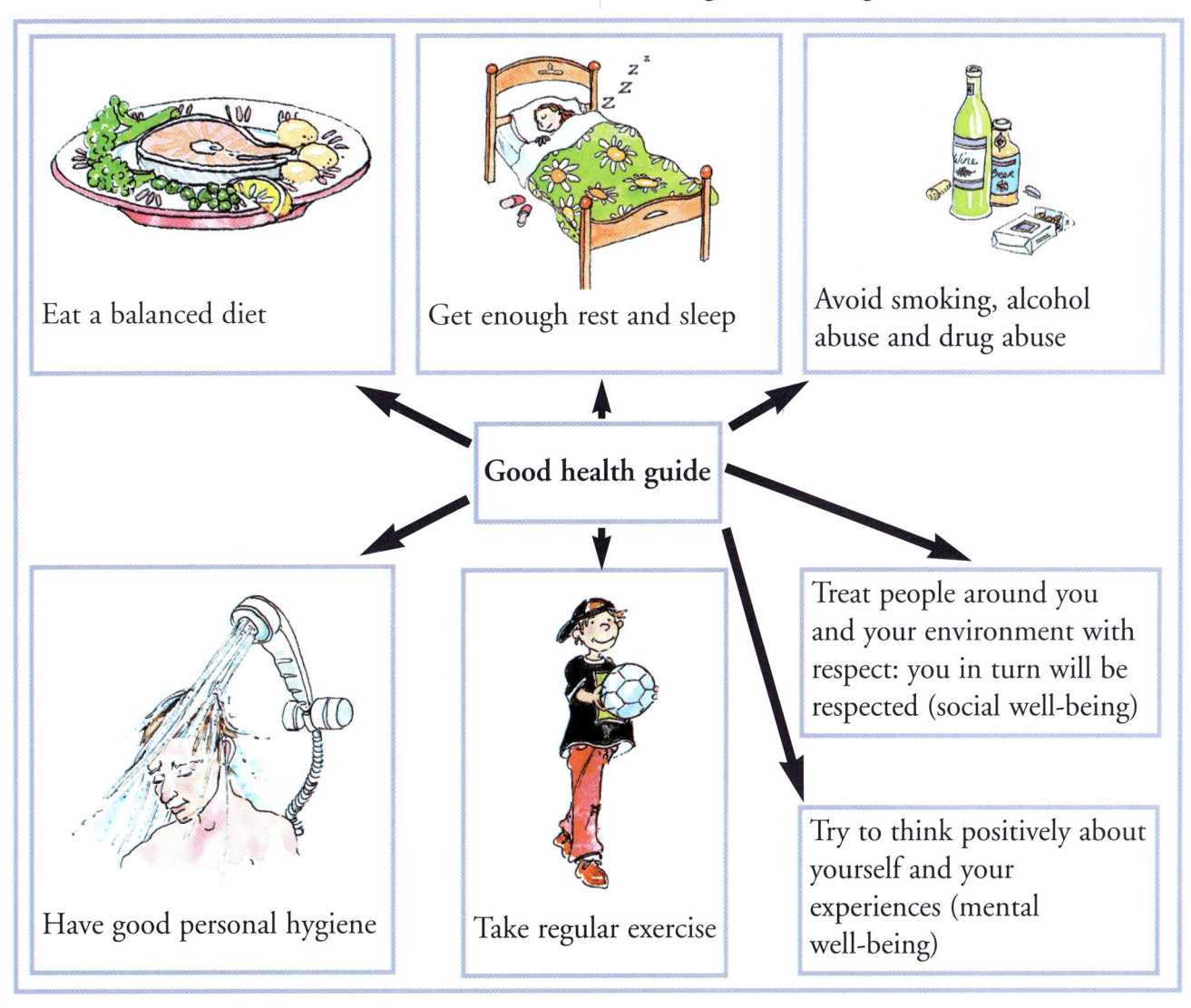

Eat a balanced diet

Get enough rest and sleep

Avoid smoking, alcohol abuse and drug abuse

Good health guide

Have good personal hygiene

Take regular exercise

Treat people around you and your environment with respect: you in turn will be respected (social well-being)

Try to think positively about yourself and your experiences (mental well-being)

ESSENTIALS FOR LIVING

Healthy diet
(See Chapter 1)
Try to follow a balanced diet; remember healthy eating guidelines:
- eat less food that is high in fat, salt or sugar, e.g. crisps
- increase fibre: eat more fruit, vegetables, brown bread, rice etc.
- drink plenty of water

Exercise
Exercise is vital to good health. Most children do not need to be told to take exercise; they just do it automatically.

Unfortunately during the late teens and early twenties many people give up sport and regular exercise. This results in unfitness and weight gain. Habits formed now can stay for life.

Activity 24.1 – Workbook p. 198
Using the excercise questionnaire in your workbook (which you may photocopy), find out how young people in your school spend their leisure time and how much excercise they take. Why not present your findings as bar charts etc.?

Relaxation and sleep
Getting enough sleep is vital to health. For a good night's sleep avoid having large meals, tea or coffee late in the evening. Stop doing homework or studying at least one hour before bedtime, as your body and mind need to be relaxed for you to sleep well.

Stress is the most common cause of sleeplessness. It is better to treat the stress itself rather than the sleeplessness alone (for example with sleeping pills, which can be habit-forming).

- reduces stress
- keeps heart and lungs healthy
- keeps weight off
- maintains good muscle tone and bone density

Benefits of exercise

204

HEALTH EDUCATION 24

Activity 24.2 – Workbook p. 199

Leisure time

Leisure time is time spent away from work, school or study. Some people waste their leisure time and then complain about being bored. They say things like 'There is absolutely nothing to do in this area or town.' This is rarely true. Sometimes leisure facilities are there but people either don't know about them or are unwilling to use them.

Activity 24.3 – Workbook p. 199

Aerobic exercise, such as jogging, step-aerobics or playing active sports, works the heart and lungs making them stronger and healthier.

Mental health

A mentally healthy person generally:
- feels positive about themselves (positive self-image or high self-esteem)
- feels positive about the people and world around them
- is able to cope with life most of the time

Mental and physical health rely on each other: neglect one and the other suffers.

Mental health guidelines

1. Look after physical health (good diet, exercise, relaxation and sleep)
2. Do not smoke or use drugs; keep alcohol within healthy limits
3. Think positively about yourself (see Self-esteem below)
4. Form a few close, trusting friendships; do not neglect these friendships, for instance if you have a new boy- or girl-friend
5. Discuss problems and feelings with a trusted friend
6. Have good routine in your life including regular sleep and meals
7. Make good use of leisure time
8. Avoid too much stress

Activity 24.4 – Workbook p. 200

ESSENTIALS FOR LIVING

Self-esteem

People who like themselves and have confidence in their abilities have *high self-esteem*. Those who do not think much of themselves or their abilities have *low self-esteem*.

Why do some people have low self-esteem?

During childhood how others treat us plays an important part in the development of self-esteem. Children who are neglected, abused, criticised or live with family problems such as alcohol abuse or poverty are more likely to have low self-esteem. In adulthood how we treat ourselves has a lot to do with self-esteem.

The issues below are sometimes linked with low self-esteem.

- alcohol or drug abuse
- smoking
- teen pregnancy
- staying in abusive relationships
- bullying (bullies usually have low self-esteem)
- possessiveness
- leaving school early
- mental illness e.g. depression
- being very critical of others
- overeating

Stress

Stress is a normal reaction to difficulties in our lives. Some stress is healthy or else we would become bored. It is when we experience too much stress that problems occur.

Common effects of stress on the mind and body

Body	Mind
Short term 　increased heartbeat 　faster breathing 　indigestion 　sweating **Long term** 　frequent headaches 　back ache 　stomach ulcers 　high blood pressure 　prone to infection 　skin problems, e.g. eczema	low self-esteem tension anxiety being easily irritated frustration anger tiredness feeling depressed tearfulness feeling unable to cope forgetfulness

Coping with stress

There are two basic ways of coping with stress:

1. Get rid of the cause of the stress (e.g. reduce your work load), or
2. Do something to relieve the symptoms:
   - eat well
   - take physical exercise
   - take time to relax
   - get enough sleep
   - talk over your problems with someone you trust

Activity 24.5 – Workbook p. 201

Exam time – Workbook p. 201 – Health

HEALTH HAZARDS

Every year in Ireland thousands of people die prematurely because they expose themselves to various health hazards. In this section we shall examine each of the following, as these are the main health hazards that young people are exposed to:

- cigarettes
- alcohol
- cannabis
- ecstasy
- solvents

Drug abuse is one of the most common health hazards. A drug is a substance that changes the way the body works or how a person acts, feels or thinks. Drug abuse is when the use of a drug harms the individual or society in some way.

ESSENTIALS FOR LIVING

Classification of drugs
There are four basic types of drugs.
1. Controlled drugs: available only on prescription e.g. sleeping pills
2. Legal drugs: alcohol, nicotine, paracetamol
3. Unrecognised drugs: glue, aerosols
4. Illegal drugs: cannabis, ecstasy, LSD, heroin

Smoking

Emphysema
Emphysema is a serious, incurable disease of the lungs. Someone with emphysema has great difficulty in breathing or doing anything that needs energy, such as climbing the stairs.

Passive smoking
Passive smoking means inhaling other people's smoke. Frequent passive smokers run the same health risks as smokers.

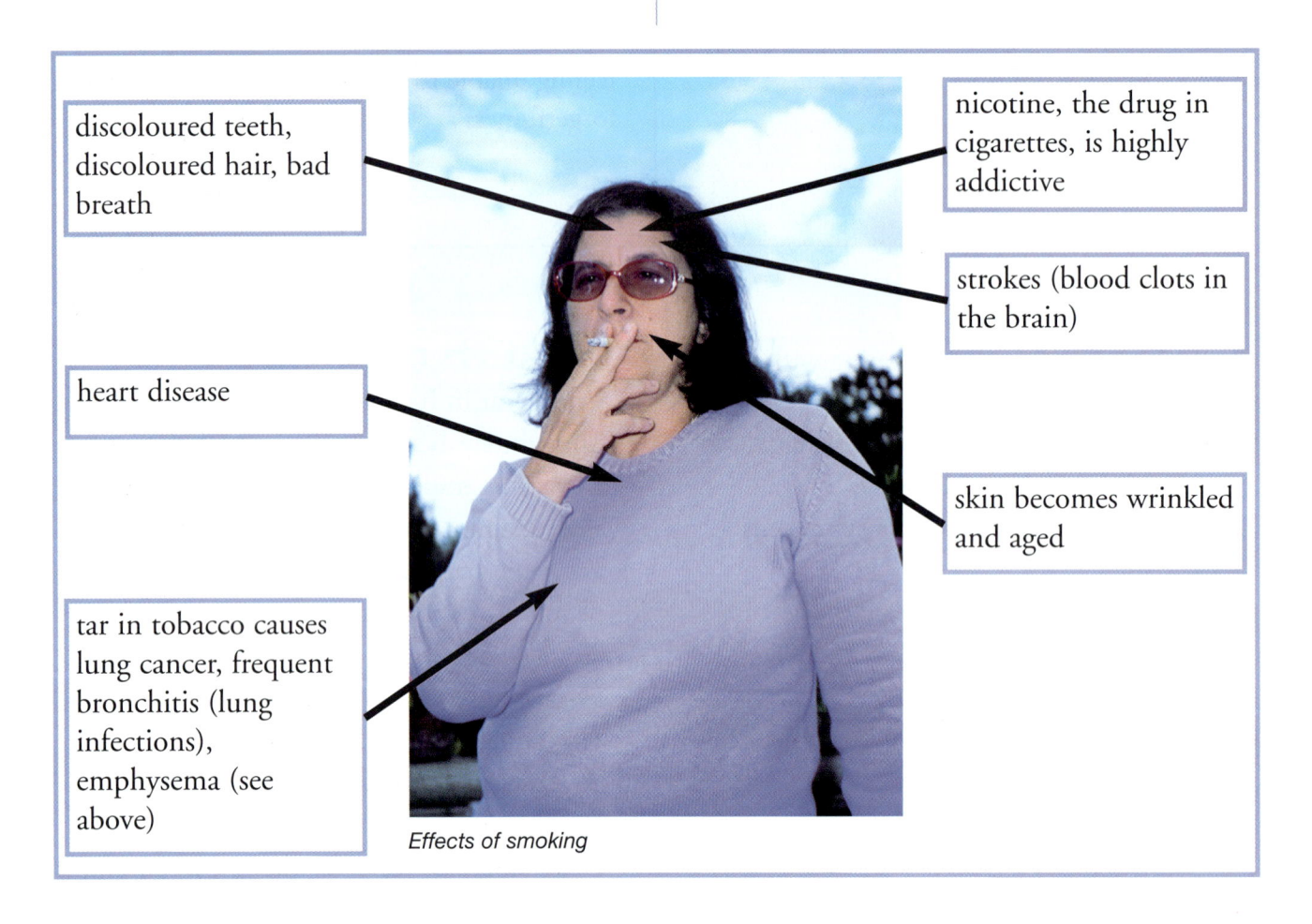

- discoloured teeth, discoloured hair, bad breath
- heart disease
- tar in tobacco causes lung cancer, frequent bronchitis (lung infections), emphysema (see above)
- nicotine, the drug in cigarettes, is highly addictive
- strokes (blood clots in the brain)
- skin becomes wrinkled and aged

Effects of smoking

Smoking during pregnancy

If you smoke your baby smokes too.
- there is an increased risk of the baby miscarrying or being stillborn
- babies of smokers are smaller and more likely to be premature
- smoking near a baby either before or after it is born increases the risk of cot death
- the children of smokers are more likely to smoke themselves

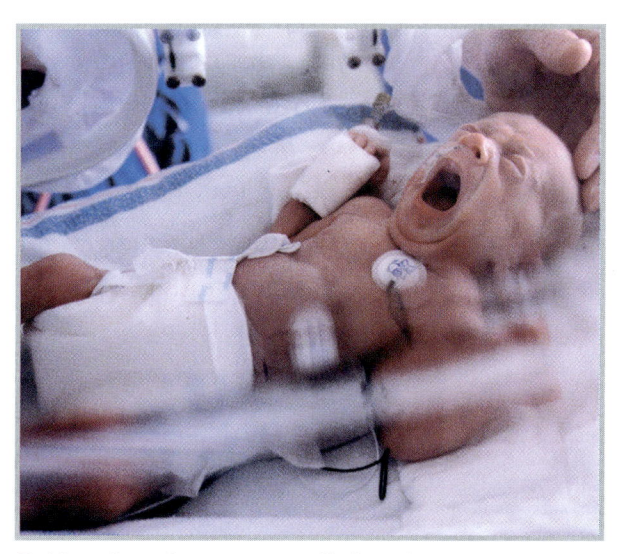

Babies of smokers are more likely to be premature

It is widely believed that giving up smoking is as difficult as giving up heroin. So why do so many young people start each year? Curiosity? Image? Peer pressure? What do you think?

Alcohol

Alcohol, when taken in moderation, can be a very positive social substance. Historically, however, the Irish have had a very unusual relationship with alcohol. We have tended to either not drink at all or else drink to excess. In fact it is estimated that one in ten Irish adults are problem drinkers (previously termed alcoholics).

Moderate social drinking

A person is considered a problem drinker when alcohol is causing a problem in their home, at school, in their social life or at work; for example missing a day from work because of a hangover.

> **Safe alcohol limits spread over the week**
> Fully grown men: 21 units per week
> Fully grown women: 14 units per week
> One unit = half a pint of beer or lager or a small measure of spirits or a glass of wine

ESSENTIALS FOR LIVING

Effects of alcohol abuse on mind and body

Short term
- face goes red
- loss of inhibitions, e.g. has unprotected sex
- mood alters: drinker may become aggressive, sad etc.
- vomiting
- lack of co-ordination
- staggering
- poor driving skills: a seventeen-year-old male driver is forty times more likely to have an accident after one and a half to two units of alcohol

Long term
- addiction
- brain damage
- mental illness, e.g. depression
- heart disease
- liver disease (cirrhosis)
- cancer of the mouth or stomach
- foetal alcohol syndrome (mental handicap caused by drinking during pregnancy)

Effects of alcohol abuse on the family

- general unhappiness and uncertainty in the family: 'Will he/she be drunk tonight?'
- separation and divorce
- violence in the home
- financial problems

Alcohol is often a factor in child abuse cases

210

HEALTH EDUCATION 24

Effects of parental alcohol abuse

Children living with parent(s) who are problem drinkers have a lot to cope with. The drinker can be very inconsistent – one minute the loving parent, the next the selfish drinker, perhaps being verbally or physically abusive.

Studies of children affected by parental alcohol abuse show that they often take on one of four roles or personalities in an effort to cope.

Hero	Scapegoat
Very mature, responsible. Works hard at school. Covers up for the drinker, tries to hold the family together in an effort to show the world there is nothing wrong in their family.	Gets into trouble at school. During teenage years: rule-breaking, e.g. alcohol and drug abuse, underage sex, teenage pregnancy. They are trying to take attention away from the drinker.
Lost child	Mascot
Quiet and undemanding. Sees the trouble that the drinker and the scapegoat are causing, thinks that if they become practically invisible this will help the family's problems to go away.	Usually the youngest child. He/she often treated like a baby. Told there is nothing wrong – becomes confused.

Activity 24.6 – Workbook p. 204

Help for those affected by alcohol abuse: head offices

Alcoholics Anonymous (help for the problem drinker themselves) (01) 453 8998

Al Anon (help for the families of problem drinkers) (01) 873 2699

Al Ateen (help for the teenage children of problem drinkers) (01) 873 2699

Without help the hero:	Without help the scapegoat:
The hero as an adult often marries someone who has a drink problem whom they can take care of – this is the life they know.	The scapegoat as an adult is prone to addiction themselves.
Without help the lost child:	Without help the mascot:
The lost child finds it difficult to form close, loving relationships.	The mascot tries to remain a child and does not like being responsible.

ESSENTIALS FOR LIVING

Effects of alcohol abuse on society

- road accidents: alcohol is a factor in forty per cent of road accidents (National Roads Authority 2002)
- crime: some people become violent when drunk
- cost of treating people for alcohol-related health problems
- absenteeism: people not turning up for work because of alcohol abuse

Activity 24.7 – Workbook p. 204

Cannabis

Cannabis, the world's most popular illegal drug, is smuggled into this country from South America, Africa, the Middle East and the Far East. Cannabis can be bought in three forms:

- cannabis herb: the leaves and flower of the plant
- cannabis resin: leaves mixed with oil and pressed into blocks
- cannabis oil: rare in Ireland

Cannabis herb is called *grass*, *weed* or *dope*. Cannabis resin is called *hash* or *blow*.

HEALTH EDUCATION   24

Effects of cannabis use

Short term
The cannabis user may:
- relax and talk more
- have a feeling of slowed time
- suffer headaches and confusion
- get a dry mouth
- vomit
- lack co-ordination

Long term
The cannabis user may:
- lose interest and ambition
- suffer memory loss and reduced ability to learn
- be absent from work or school
- get lung cancer (the risk is three times higher than the risk from cigarettes)
- suffer mental illness, e.g. depression
- go on to use other drugs
- if caught, get a criminal record

Ecstasy

Ecstasy was invented in Germany in 1910 as a slimming tablet. The drug was never made legal because of its dangerous side effects. 'E' takes the form of a small white tablet with a logo on it. Logos include doves, shamrocks and Mitsubishi signs.

After taking ecstasy the user becomes full of energy: their heart begins to beat faster and blood pressure rises. The user begins to sweat a lot and becomes thirsty.

Ecstasy became popular in the 1980s and 1990s with the coming of the dance culture

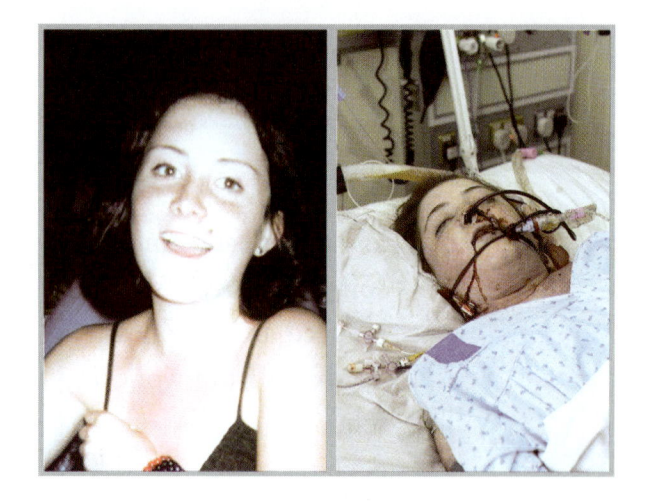

Lorna did not die because of a car accident. She did not have a fall. On the right is Lorna after taking two 'E' tablets. Her father and mother said they hoped that this picture would serve to warn others. Her heartbroken mother said that her beautiful daughter died looking like a monster. Police said she had taken two lime-coloured pills marked with the Euro symbol before visiting a nightclub in Cambridge where she was studying sociology. (LCA Social Education Exam 2002)

Solvent abuse
Solvent abuse means deliberately inhaling the fumes from substances such as glue, aerosols, petrol etc.
- in Ireland up to ten people die from solvent abuse each year
- solvent abusers appear drunk
- solvent abusers sometimes hallucinate, have headaches, lose their appetite or are unable to concentrate
- frequent solvent abusers have a rash around their mouth

Other drugs common in our society are heroin, cocaine, speed, LSD and prescribed drugs that are abused, such as valium.

Health promotion
Most countries ban or limit the use of substances that are hazardous to health. There are laws banning drink driving and some drugs. Health promotion tactics include health warnings on cigarettes and drug education in schools.

For further information on any health issue contact:
The Health Promotion Unit,
Hawkins House,
Dublin 2
Phone: (01) 670 7987
They will give you the address and number of the Health Promotion Unit for your area.

HEALTH EDUCATION

Effects of drug abuse on the individual

- health problems, e.g. HIV, from sharing dirty needles
- physical addiction: the body craves the drug. Heroin is physically addictive
- psychological addiction: relying on the drug to cope with life
- loss of self-respect: an addict will do anything to get the drug
- mental illness, e.g. depression, paranoia

Effects of drug abuse on society

- crime: many addicts steal to feed their habits
- cost to the tax payer
- healthcare for addicts
- unemployment benefit for addicts unfit to work

Activity 24.8 – Workbook p. 206

Exam time – Workbook p. 206 – Health hazards

THE SKIN AND PERSONAL HYGIENE

Excretion means getting rid of waste products. There are a number of different parts of the body that excrete waste: these are called excretory organs. They include:

Organ	Waste product
lungs	CO_2
kidneys	urine
intestines	faeces
skin	sweat

The skin

The skin has two layers:

- the epidermis (the surface)
- the dermis (under the surface)

The epidermis is the outermost layer of the skin. As you can see in the diagram the top layer of the epidermis consists of flat, dead cells. These cells are being constantly brushed off and replaced from beneath. The *Malpighian layer* lies further into the epidermis. This layer contains a pigment (colour) called melanin. The more melanin you have, the darker your skin is. Melanin protects the skin from the sun.

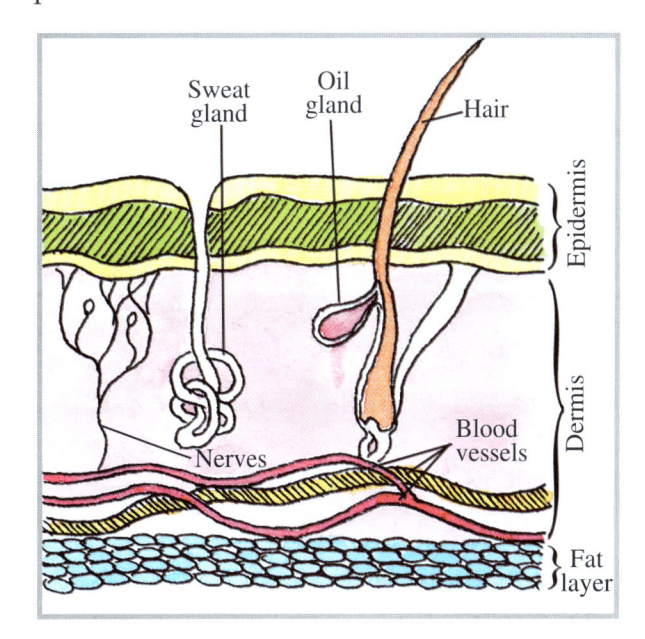

The dermis is under the epidermis and contains:

- blood vessels
- nerves
- sweat glands
- oil glands
- hair
- a fat layer

Each of these has a function.

ESSENTIALS FOR LIVING

Functions of the skin

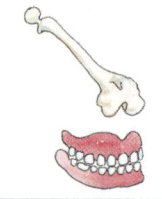

1. **Vitamin D**
   Vitamin D is made by the skin from sunlight. It is needed for healthy bones and teeth

2. **Protection**
   The epidermis stops bacteria getting into the body and also prevents loss of body fluids (burn victims are prone to infection and dehydration)

3. **Temperature control**
   - blood vessels in the skin widen when we are hot
   - the skin goes red thus giving off unwanted heat
   - sweat glands produce sweat which cools us down

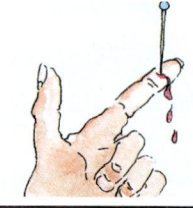

4. **Organ of touch**
   Nerves in the skin allow us to feel pain, cold, heat etc.

5. **Insulation/energy store**
   The fat layer keeps us warm and acts as an energy store

Personal hygiene

Body odour (BO): Even while at rest our sweat glands are constantly producing sweat (about one litre per day). When this sweat dries it mixes with oil from the oil glands and sits on the surface of the skin. If it is not washed off body odour (smell) is the result. During the teenage years sweat and oil glands become very active and teenagers are much more prone to body odour than children. To avoid body odour:

- eat plenty of fruit and vegetables
- drink plenty of water
- avoid cigarettes and alcohol
- get enough sleep
- don't sleep with make-up on
- wash daily

HEALTH EDUCATION 24

Skin hygiene routine

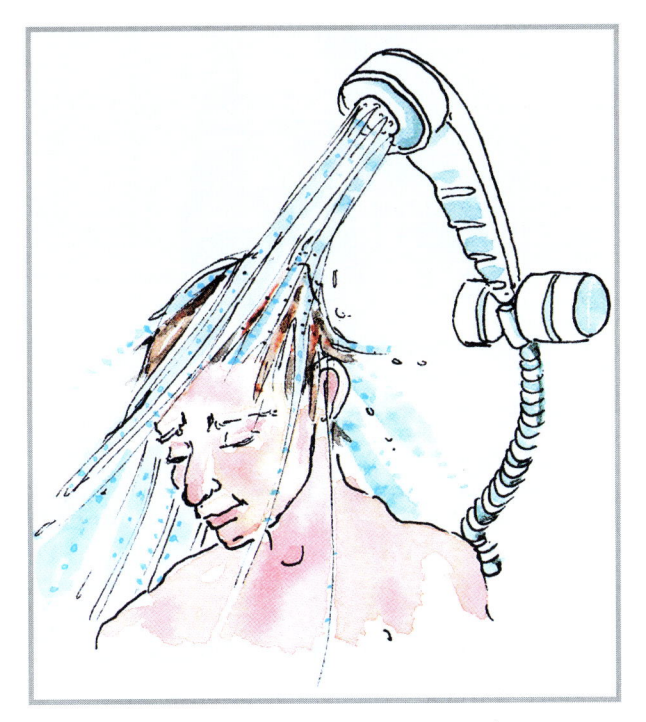

1. Shower every day, use soap or gel.

2. Use deodorant or better still anti-perspirant.

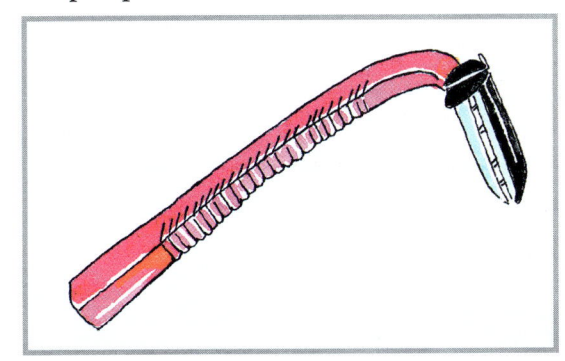

3. Girls: shave underarm hair: it traps BO.

4. Shower after sports.

5. Change socks and underwear daily.

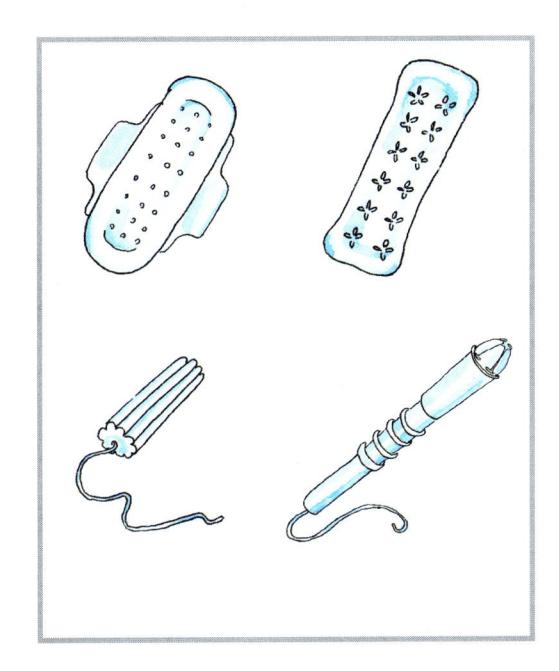

6. Girls: be particularly careful during periods, as BO can be a problem at this time.

ESSENTIALS FOR LIVING

Acne

Acne is a skin condition that usually starts during the teenage years. A teenager's skin produces extra oil, which can block pores and cause blackheads and spots. Acne cannot be prevented but it may be helped by:

- avoiding fatty foods e.g. chocolate and crisps
- not squeezing spots (spreads acne)
- drinking plenty of water
- using medicated soap, e.g. Biactol

Acne is a problem experienced by many teenagers

Skin cancer

In Ireland 5,500 people per year are diagnosed with skin cancer (Irish Cancer Society 2002).

Reduce the risk by:

- avoiding too much sun
- wearing sun cream

Children's skin is easily damaged; they must wear a high factor sun cream.

Now test yourself at *www.my-etest.com*

Sunbathing = skin cancer and wrinkles

Care of the hair, hands and feet

Care of the hair	Care of the hands and feet
• wash hair at least once a week • use conditioner on the ends • rinse very well to prevent dandruff • avoid too many hair treatments or blow drying • trim hair every six weeks	*Hands* • wash hands, especially after using the toilet • keep nails clean and filed • use hand cream • use rubber gloves *Feet* • wash daily (dry well between the toes) • wear good supportive footwear • cut nails straight across • change socks daily

Revision crossword – Workbook p. 207

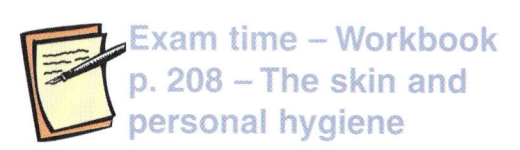

Exam time – Workbook p. 208 – The skin and personal hygiene

Unit 4

Resource Management and Home Studies

chapter 25

Home management

Good management means using *resources* wisely to achieve *goals*.

RESOURCES

Resources are what we use to carry out tasks (common tasks include making dinner and cleaning the house). There are four basic resources:

- time
- human energy
- skill
- money (and what money can buy)

Human energy

Time

Skill

HOME MANAGEMENT | 25

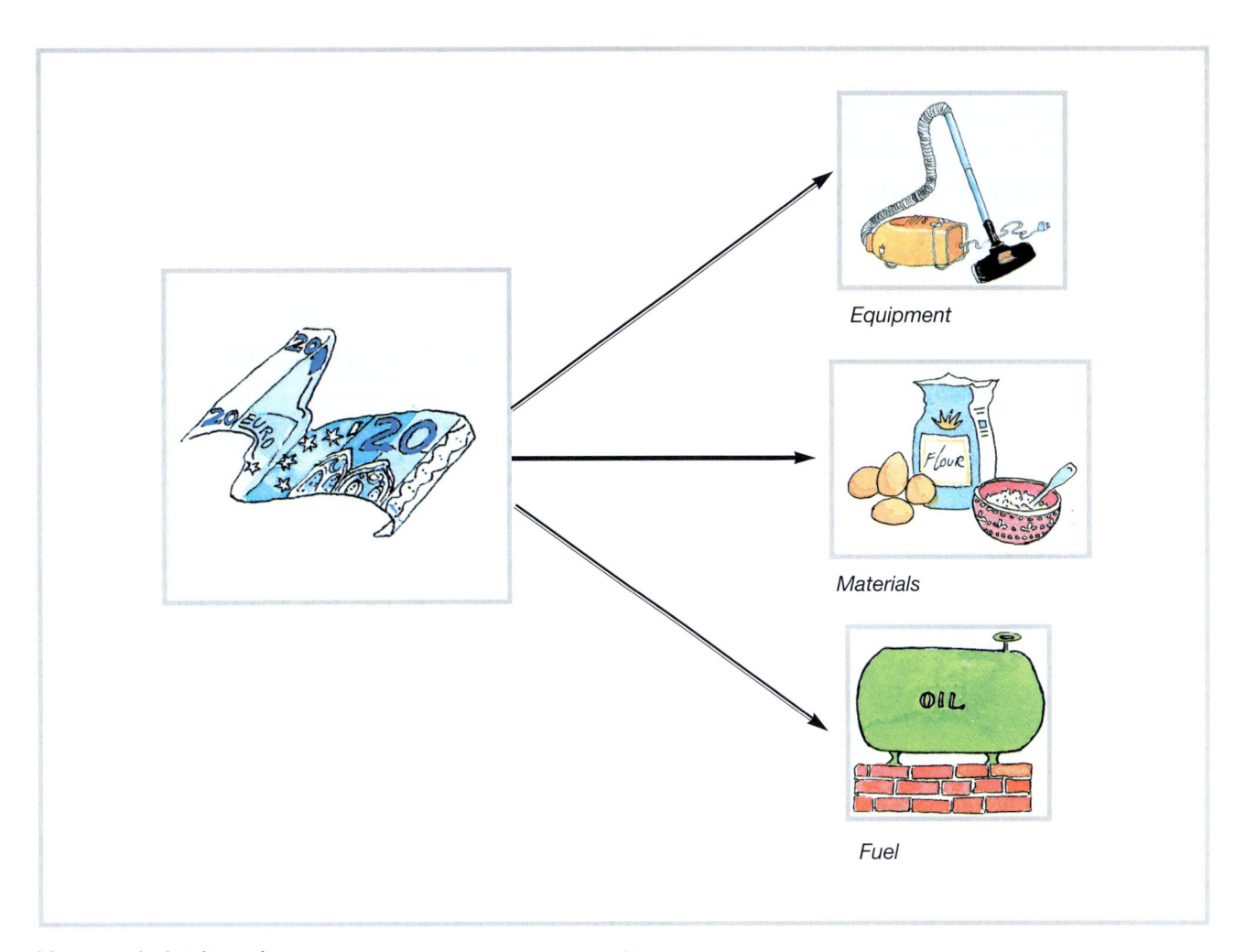

Money and what it can buy

Equipment

Materials

Fuel

GOALS

Goals are things that you or your family want to do or achieve. Goals can be:

- short term, for example to make the dinner
- medium term, for example to keep the house tidy this week
- long term, for example bit by bit to redecorate the house

A GOOD MANAGEMENT SYSTEM

- name your goal, e.g. Clean my room
- list resources needed, e.g. cleaning agents, duster, cloths, vacuum cleaner
- make a work/time plan (see below)
- carry out action: clean the room
- evaluate: Is my room clean? What did I do well/badly? Quickly/slowly? Any changes next time?

Work/time plan

(see page 64 for cookery time plans)
A work/time plan is list of smaller tasks placed in a logical order, so as to achieve a larger goal.

> ### Work/time plan: spaghetti bolognese
> 10.00 Gather all ingredients and equipment
> 10.10 Chop onion, crush garlic
> 10.20 Fry mince, onion and garlic
> 10.30 Add tomatoes, stock, salt and pepper
> 10.35 Bring to boil, reduce heat, simmer for 30 minutes
> 10.45 Cook spaghetti
> 11.05 Serve, wash up

Running a home

The main jobs involved in running a home are:
- budgeting
- shopping
- cooking
- cleaning and household maintenance
- caring for children
- washing and ironing clothes

While some of the jobs listed, for example budgeting, are usually done by parents, others, such as cleaning, can be done by all family members, even the very young.

Weekly cleaning roster/work routine

1. List all the jobs that need to be done and divide them up according to how often they have to be done.
   - daily, e.g. wash dinner dishes
   - weekly, e.g. vacuum stairs
   - occasionally, e.g. wash windows
2. Divide out the jobs according to how much free time the person has and how old they are (for example a young child couldn't be expected to cook dinner).
3. Make sure two or three weekly jobs are done every day and that one 'occasional' job is done every week.
4. Try out the roster; change if necessary.

✏️ **Activity 25.1 – Workbook p. 213**

Cleaning a room
- tidy up
- do messy jobs first, e.g. clean out the fire
- sweep (if not vacuuming)
- dust
- vacuum
- wash, e.g. skirting boards
- polish, e.g. furniture

✏️ **Activity 25.2 – Workbook p. 215**

chapter 26

Home design

Shelter is a basic human need. Even though house prices in Ireland remain high, a large percentage of Irish people are home owners.

Types of accommodation
Houses: bungalow, two to three storey, detached, semi-detached, terraced
Apartments: purpose-built or a large house divided into flats or bedsits (one room)
Institutions: convents, hostels (such as Homeless Aid), boarding schools
Sheltered housing: a number of small housing units built together, for example for elderly or disabled people; there is a warden and sometimes a doctor or nurse on call
Caravan, mobile home: used by the Irish travelling community. Young people often live in mobile homes on site while their house is being built to save spending money on rent

Sheltered housing

Making a house a home
A house simply provides shelter; a home contains people and provides for other needs as well.
Physical needs: shelter, warmth and protection
Emotional needs: safety, security, love, privacy (curtains are one of the first things people buy for a new home)
Social needs: it is at home that we learn our basic social skills, for example table manners, as children. People feel comfortable and can be themselves at home.

ESSENTIALS FOR LIVING

DESIGN

Good design is when something is (a) functional, (b) attractive to look at and (c) hard-wearing.

> **Features of design**
> 1. Function
> 2. Form (shape and line)
> 3. Colour
> 4. Pattern
> 5. Texture

1. Function

A good design must be able to serve its function. For example a chair must be strong enough to hold a person and be comfortable to sit on.

2. Form (shape and line)

Shape

Most objects in the home, rooms and even houses themselves use four basic shapes: square, rectangle, triangle and circle.

What shapes do you see in this traditional-style kitchen?

Line

Lines in rooms, houses, furniture or objects can be vertical, horizontal or curved. Each creates a different effect.

Vertical lines cause the eye to look up and down. They make things look taller and narrower.

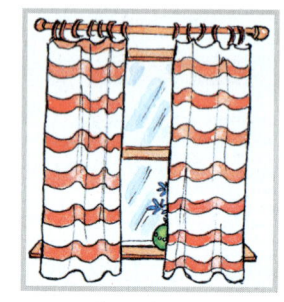

Horizontal lines make things look shorter and wider. Which window is tallest? Which is widest?

Curved lines appear soft and feminine.

224

HOME DESIGN | 26

3. Colour

Colour is a very important element of design, especially interior design. Paint a room white and you will create a feeling of light and space. Paint that same room plum and it will feel smaller and cosier.

The colour wheel

Primary colours: blue, yellow, red
Secondary colours: two primary colours mixed
Tertiary colours: one primary colour and one secondary colour mixed, e.g. blue and green = turquoise

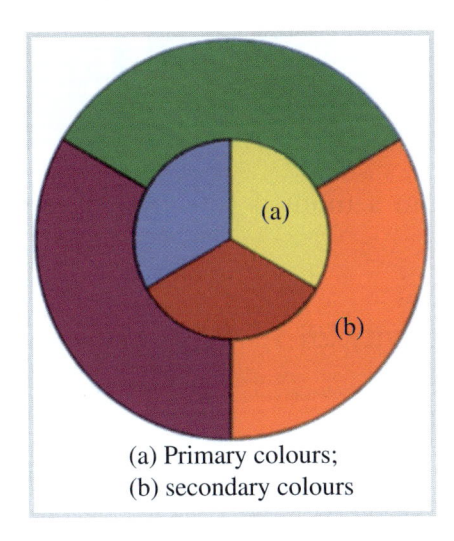

(a) Primary colours;
(b) secondary colours

Describing colours

Colours can be described as being (a) warm, (b) cool, (c) neutral or (d) pastel. Each of these can be used to create a different effect in the home.

(A) Warm colours: vibrant warm colours such as reds, oranges, pinks and purples are very rich. When used alone they make a room look much smaller than it is. They are best used in small amounts with neutrals or pastels. For example, paint three walls white with one wall a rich, warm colour.

Warm colour adds interest to pastels and neutrals

(B) Cool colours like blue and green create an atmosphere that is cool and restful. Best used in warm sunny rooms.

225

ESSENTIALS FOR LIVING

(C) Pastels like light blue, green, pink and yellow are also restful and great for bedrooms.

(D) Neutral colours: black, white, cream and beige are often called neutral colours. Interest can be added to neutral colours by adding splashes of strong colour.

A shade: black is added to a colour to darken it
A tint: white is added to a colour to lighten it

4. Pattern

Pattern can add interest to a room. Too much pattern makes a room look crowded and untidy. It is possible however to have several different patterns in a room if you keep their colour the same.

✏️ **Activity 26.1 – Workbook p. 216**

5. Texture

Texture describes how an object feels: rough, smooth, warm or cool.

Smooth texture, for example the texture of

226

tiles and painted surfaces, gives a cool, clean, hygienic feel.

Rough texture, for example the texture of carpets and curtains, absorbs sound and gives a warm, cosy feel.

Higher level

In design there are four other important elements. These apply well to interior design.

i) **Proportion**: This relates to the size of objects. Pieces of furniture in a room should be *in proportion* to each other and to the room. For example a large antique table would be *out of proportion* in a small apartment.

ii) **Emphasis**: When the eye is drawn towards something in the room. This is called the *focal point*. Example: a fireplace.

iii) **Balance**: When there is an equal spread of colour, pattern, and texture in a room.

iv) **Rhythm**: This is where a colour, shape or pattern links or ties a room together. Look at the room in the picture.

An example of rhythm: what colour is tying this room together?

Exam time – Workbook p. 216

Now test yourself at *www.my-etest.com*

chapter 27

Room planning

When planning or redecorating a room, consider each of the following:

1. **Function**: Which room are you planning? If it is a sitting room it needs to be comfortable and relaxing. A kitchen, on the other hand, needs to be hygienic and easy to work in.
2. **Existing fixtures and fittings**: Doors, windows, fireplaces and radiators cannot easily be moved and need to be considered in a plan.
3. **Heating and lighting**: This must be decided upon early as rewiring etc. may need to be done.
4. **Storage**: A room with little storage will be difficult to keep tidy. Plan plenty of storage space.
5. **Traffic flow**: Furniture should be placed so that people can walk around the room without bumping into it. Do not put too much furniture into a room.
6. **Colour**: The colour you choose for a room will often depend on its aspect, that is whether it is north or south facing. North facing rooms tend to get less sun so warm colours are often used. South facing rooms get lots of sun so cool colours can be used.
7. **Pattern and texture**: The use of pattern and texture can add interest to a room.

Floor plan

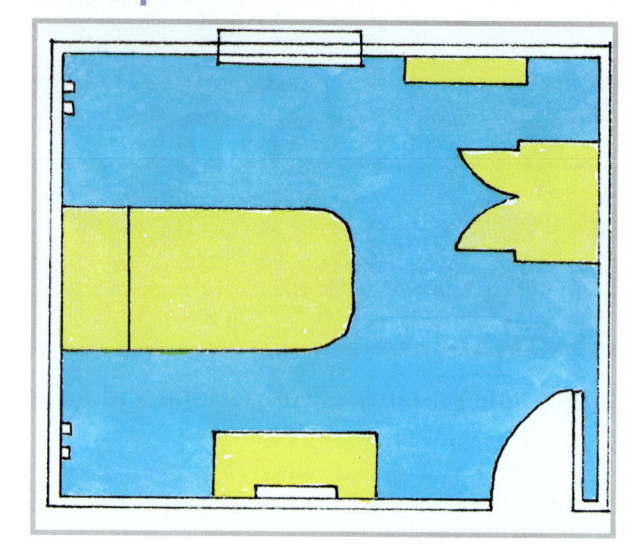

A floor plan

When planning a room it is wise to make a floor plan. The room outline should be drawn *to scale* on graph paper. Fixtures such as doors and windows are then drawn in. Furniture can be drawn to scale, cut out and then moved around on the floor plan until the best arrangement is achieved.

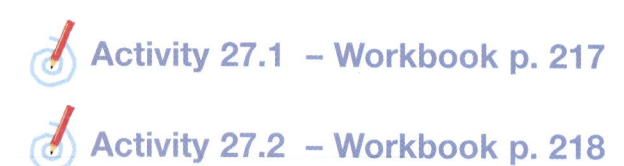

✏️ Activity 27.1 – Workbook p. 217

✏️ Activity 27.2 – Workbook p. 218

Kitchen design

A Kitchen must:
- be hygienic and easy to keep clean
- be efficient: not too much walking between the fridge, sink and cooker
- have enough storage space
- be bright and well-ventilated
- be safe to work in

Ergonomics

Higher level
Ergonomics is the study of how efficiently people work. Design has a lot to do with ergonomics. In a badly designed workplace frequently used pieces of equipment will not be conveniently located.

The work triangle
The fridge, sink and cooker are the three pieces of equipment used most in the kitchen. It makes sense to locate them near each other in an invisible triangle. There should be work surfaces in-between.

Too much walking between the fridge, sink and cooker

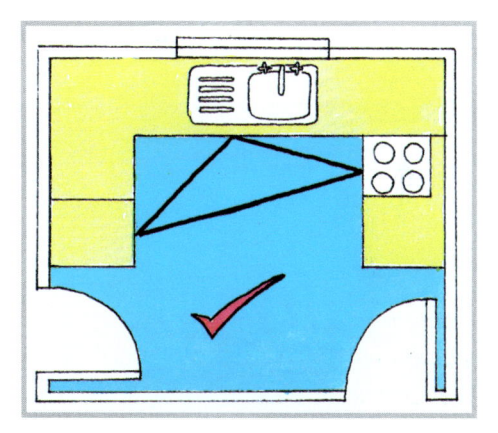

A well-designed kitchen

Exam time – Workbook p. 219

Now test yourself at *www.my-etest.com*

chapter 28

Services to the home

ELECTRICITY

Electricity is a form of energy produced and supplied in this country by the Electricity Supply Board (ESB). Electricity is produced by the movement of either wind or water or by burning gas, turf, oil or coal.

Wind and water: renewable energy sources

When the electricity leaves the ESB power station it travels in huge cables to our homes. Electricity enters the house through the *fuse box* and then goes to all the light fittings and sockets in the house.

The amount of electricity used is recorded by an *electricity meter*. Sometimes there are two meters, one for *Night Saver* electricity (night-time). Households are sent a bill every two months.

Electricity meter

Modern fuse box

Activity 28.1 – Workbook p. 223

Modern fuse boxes do not contain actual fuses, instead they contain switches called *circuit breakers*. These switches click off if there is a fault. They are easily switched on again when the fault is corrected.

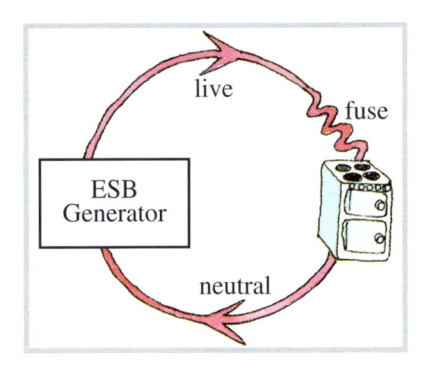

SERVICES TO THE HOME | 28

Electrical circuit

Electricity always travels in a circuit or circle. A *fuse* is a deliberate weak link in the circuit. If anything goes wrong the fuse blows and the circuit breaks. This stops the electricity flowing. Fuses can be found in the main fuse box of the house and in electric plugs.

All electrical appliances must have at least two wires:

- live (brown)
- neutral (blue)

The *live* wire brings electricity to the appliances in our homes and the *neutral* wire returns to the ESB generator. A third wire called the *earth* wire (yellow and green) is for safety. If something goes wrong electricity will flow to the ground through this earth wire and not through the person using the appliance. Appliances with no earth wire should be doubly insulated (see diagram).

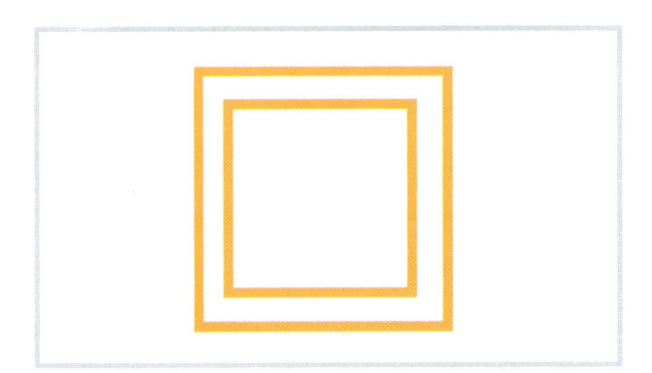

Doubly insulated

Wiring a plug

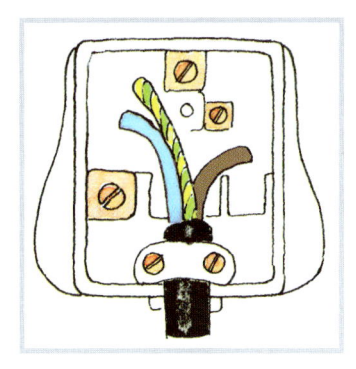

- open plug top
- clamp wire in place
- trim three small wires to size

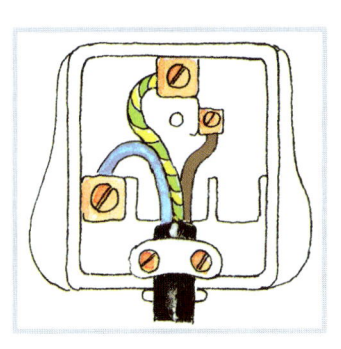

- screw each wire into correct place
- replace fuse and top of plug

brown = live
blue = neutral
green and yellow = earth

231

ESSENTIALS FOR LIVING

Electrical safety

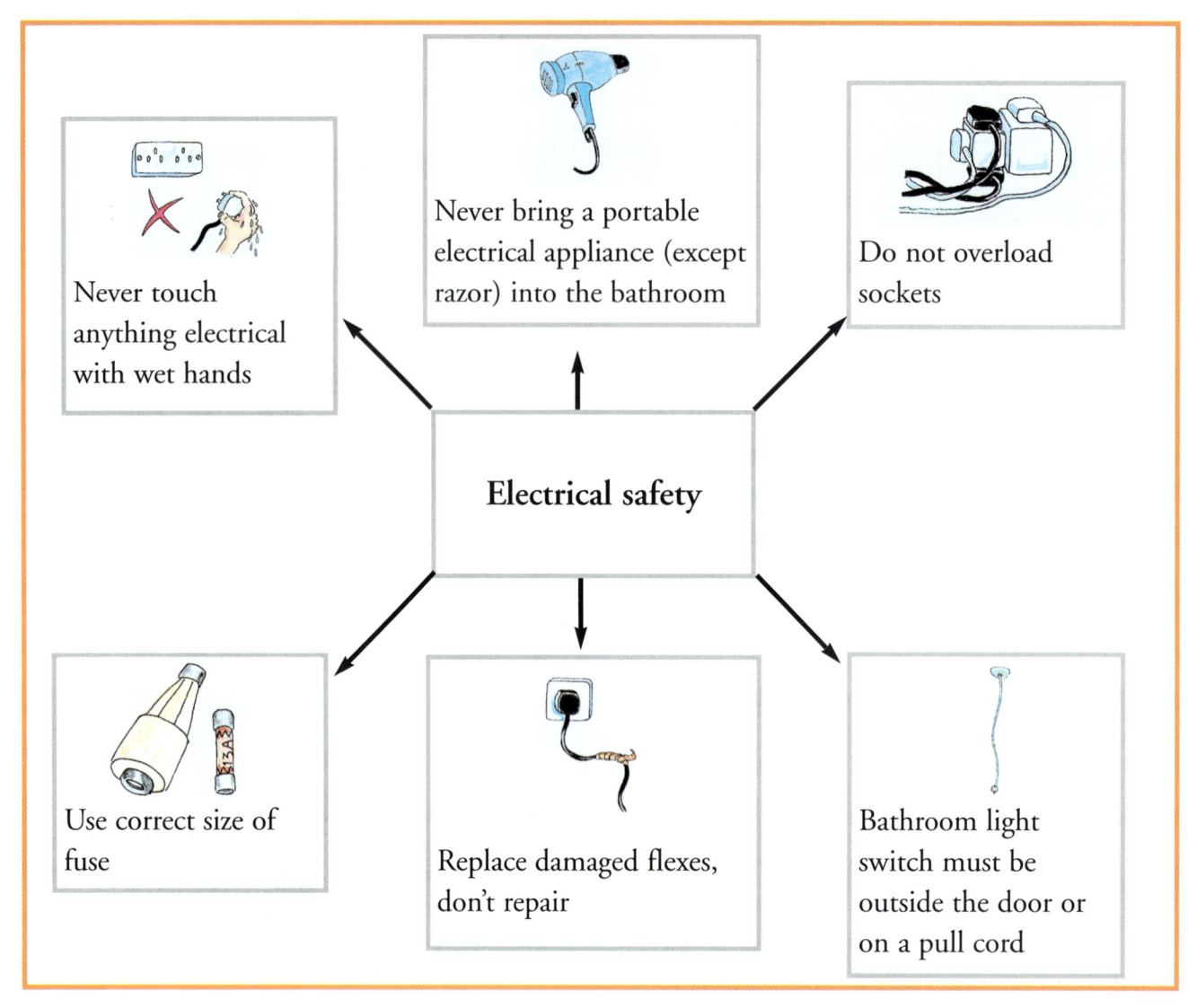

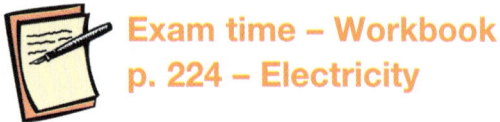

- Never touch anything electrical with wet hands
- Never bring a portable electrical appliance (except razor) into the bathroom
- Do not overload sockets
- Use correct size of fuse
- Replace damaged flexes, don't repair
- Bathroom light switch must be outside the door or on a pull cord

Exam time – Workbook p. 224 – Electricity

GAS

There are two types of gas available in Ireland.

1. Natural gas (our natural gas supply is found under the seabed at Kinsale in Co. Cork. From there it is piped to homes along the east coast up as far as Dundalk in Co. Louth).
2. Bottled gas, available in tanks or cylinders.

Natural gas fire: clean and attractive but not very warm

Gas safety

■ Gas can be dangerous: appliances should be fitted only by a qualified person.

■ Gas needs proper ventilation. Gas used in a badly ventilated room can, and does, kill.

■ If you smell gas act quickly:

Do:
- open doors and windows
- check if gas has been left on and an appliance switched off 
- if not turn gas off meter valve 
- call gas company from a neighbour's house

Don't:
- smoke or light matches 
- use anything electrical – even a light switch

Exam time – Workbook p. 225 – Gas

WATER

Water is the most basic service to our homes. Fresh water is supplied to our homes by the Corporation (urban), County Council (rural) or by a private water scheme (rural). Water is collected in a lake or a man-made reservoir and from there it goes to a treatment plant to be cleaned.

Water treatment

1. Filtered to remove impurities
2. Chlorine added to kill bacteria
3. Fluoride added for strong bones and teeth

Reservoir in Co. Wicklow

Clean water travels in a mains pipe to the houses of the area. Water enters each individual house through a service pipe. There is a valve on this pipe which can turn the water off.

Fresh water goes directly to the kitchen sink and also fills the tank in the attic. Water from this tank supplies the toilet, cold taps in the bathroom and the cylinder in the hot press.

The kitchen sink

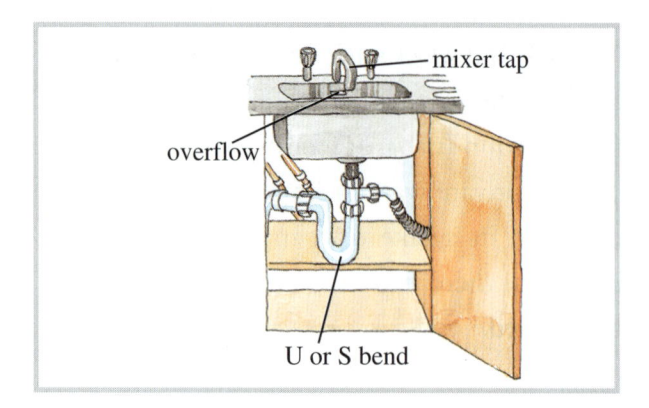

The kitchen sink is usually placed under a window. There are several reasons for this:
- light
- ventilation
- easily plumbed
- it is nice to look out the window while working

Kitchen sinks are usually made from stainless steel because it is hygienic, durable and does not stain. All sinks have an S- or U-bend. The S- or U-bend stops germs and smells from coming back up the plug hole. Sinks can become blocked by substances, especially grease.

Unblocking a sink

- Cover overflow with a cloth. Place a plunger over the plug hole and plunge up and down. If this does not work:
- Put washing soda in the plug hole and rinse down with boiling water; plunge again. If this doesn't work:
- Place a basin under the U-bend, unscrew it and remove whatever is blocking the sink with a piece of wire clothes hanger. Flush out and replace the U-bend.

Burst pipes

In the cold weather pipes may freeze. Water expands when frozen and in this way cracks the pipes. It is not until the pipes thaw again and flooding occurs that we notice a problem.

What to do:
- turn off the water at the mains
- run cold taps to clear the pipes
- turn off heating
- call a plumber

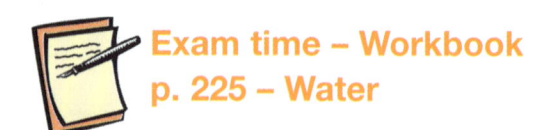

Exam time – Workbook p. 225 – Water

LIGHTING

(picture light, standard lamp, table lamp labelled on photograph)

Good lighting is needed to prevent eyestrain and accidents in the home. Lighting is also a great interior design tool and can add a special atmosphere to a room. Most rooms have a large central light. This light can be supported by other smaller ones such as table lamps, standard lamps, wall lights, spotlights etc. (see photograph).

Types of lighting

1.

Sun

2. Tungsten filament: will last for 1000 hours (approximately). The strength of the bulb is measured in watts: 100 watt bulbs are usual for a central light, 40 watt for a table lamp. Cheap to buy.

Filament bulb

3. Fluorescent tubes: glass tubes 0.5–2.5 m long. Last 3000 hours (approximately), give off a cold, clinical light.

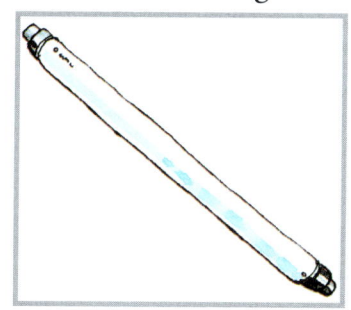

Fluorescent

4. Compact fluorescent lamps (CFLs): expensive to buy but last 8000 hours (approximately). Cheap to run, give a cold, clinical light.

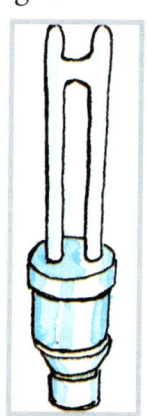

CFL

ESSENTIALS FOR LIVING

HEATING AND INSULATION

Methods of heat transfer

Heat can be transferred from a heat source such as a fire to a person in three ways:

1. **Conduction** (heat travels along something solid, the entire object becomes hot)
2. **Convection** (air or water is heated, rises, and is replaced by cool air or water – the cycle continues)
3. **Radiation** (heat rays heat the object or person they shine on)

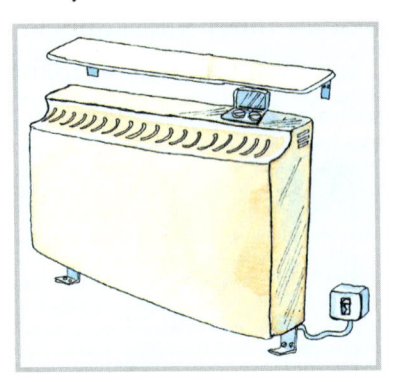

A *storage heater* heats by conduction. Heat travels through bricks in the heater, bricks become hot and begin giving off heat (by convection).

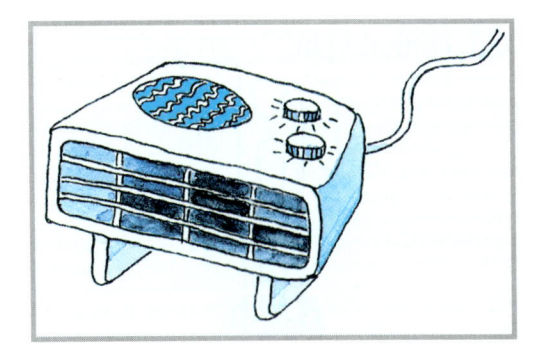

A *fan heater* heats by convection. It takes cold air in at the bottom, the element inside warms the air and gives off hot air at the top.

A *bar fire* heats by radiation. Heats whatever it shines on, not the air in-between.

A home can be heated by:

1. **Central heating**: a boiler (fuelled by gas, oil etc.) heats radiators in each room. Radiators are connected to each other and to the boiler by metal pipes.
2. **Individual heaters**: storage heaters use off-peak electricity and can be left on. Others are used for shorter periods of time, such as gas 'supersers' or open fires.

Open fire

Thermostats and timers

Thermostats and timers control when your heating system is switched on and off.

Thermostats can be set for a certain temperature, such as 21°C in a living room. Once that temperature is reached the thermostat switches the heating system off. Thermostats can be:

- on the central heating boiler itself
- on each radiator
- on the wall in (for example) the living room

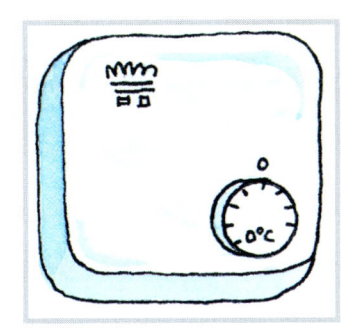

Room thermostat

Radiator thermostat

Timers can be set to switch the heating on or off at various times during the day or night, for example the heating can be set to come on two hours before you return home in the evening.

Timer

Insulation

Insulation means trapping heat in the house. A house with poor insulation will be cold, draughty and expensive to run. Windows, roofs, walls, floors and doors should all be insulated with materials that are *bad conductors of heat*. Bad conductors do not let heat pass through them easily. Examples of bad conductors are: air, fabric, fibreglass, polystyrene.

1. Attic

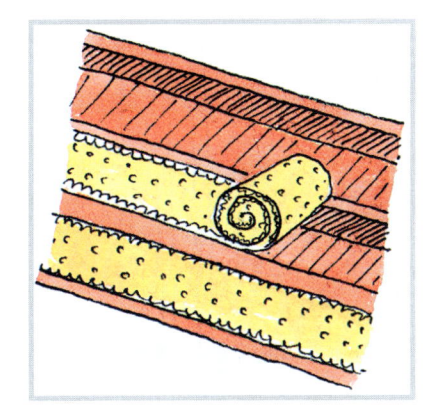

Rolls of thick fibreglass are laid on the floor of the attic to prevent heat loss through the roof.

ESSENTIALS FOR LIVING

2. Walls

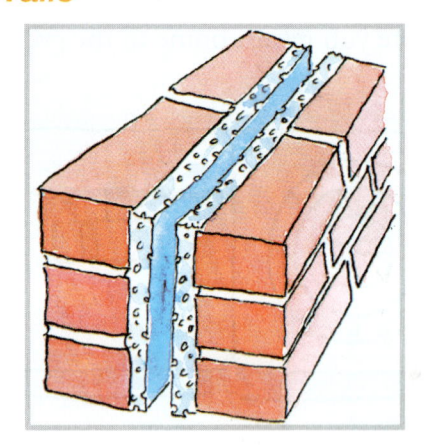

Modern houses have cavity walls. This means that there are two rows of blocks or bricks with a gap between them. Trapped air and white polystyrene sheets in the gap or cavity stop heat escaping from the house through the walls.

3. Floors

Carpet and carpet underlay are both bad conductors – good insulators.

4. Windows

Double-glazed windows help insulate the house because there is air trapped between the panes of glass. Air is a bad conductor – good insulator.

5. Draughts

Heavy curtains or draught excluders can prevent heat loss through windows, doors and letter boxes.

Exam time – Workbook p. 228 – Lighting, heating and insulation

CONSERVING RESOURCES IN THE HOME

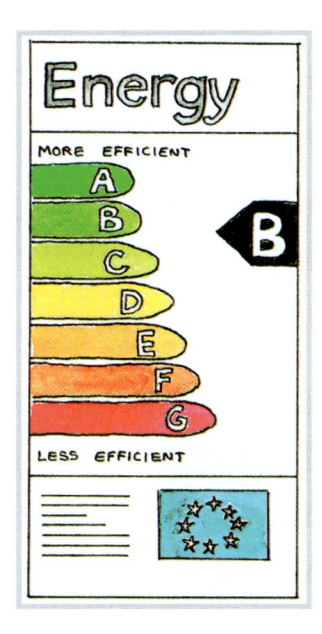

Choose appliances with a good energy efficiency rating (A or B).

Fit hot water cylinder with a lagging jacket (new cylinders come ready insulated).

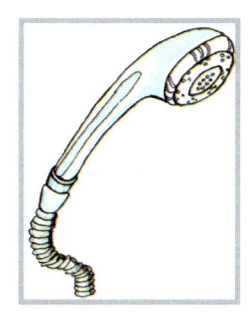

Have showers instead of baths – they use less water.

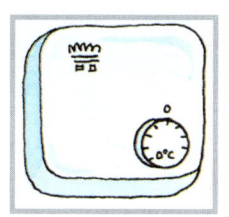

Fit thermostats and timers.

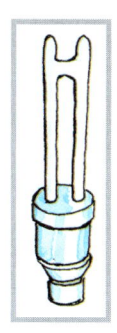

Use CFL bulbs, switch off lights when not in use.

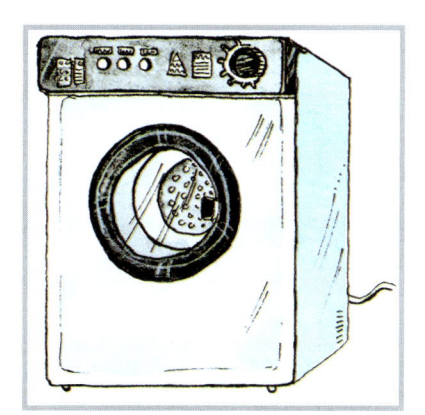

Wait for a full load before using the dishwasher or washing machine.

Insulate the home.

Now test yourself at *www.my-etest.com*

chapter 29

Technology in the home

Advances in technology have made running the home very different than it was even thirty or forty years ago.

- Washing and caring for clothes has been made much easier by automatic washing machines, tumble dryers and electric steam irons.
- Dishwashers and vacuum cleaners allow us to wash dishes and clean carpets more efficiently.
- Food preparation and storage is aided by equipment such as freezers, refrigerators, microwaves, food processors, modern cookers etc.

- The *Internet* can be used for shopping, gathering information, booking holidays, banking on-line, sending e-mail etc.

Activity 29.1 – Workbook p. 230

Modern household appliances

There are three basic types of household appliance:

1. Appliances with a motor (noisy), e.g. vacuum cleaner, electric knife, food processor

2. Appliances with a heating element, e.g. toaster, heater, kettle, cooker etc.

TECHNOLOGY IN THE HOME   29

3. Appliances with both a motor and an element (may not be a heating element), e.g. washing machine, dishwasher, fridge and freezer

Activity 29.2 – Workbook p. 230

Buying electrical appliances

- **Cost**: compare different models and the same models in different shops
- **Safety**: buy from a reliable shop; check for safety labels

European Standards mark

BSI Standards mark

Doubly insulated

- **Energy efficiency label**: appliance should ideally have an A or a B rating
- **Size**: will the appliance fit in the space you have for it?
- **Special features**: don't spend money on appliances with lots of special features if you will never use them
- **Demonstration**: if necessary ask to have a demonstration on how to use the appliance
- **Check guarantee**: it should be for at least one year
- **After-sales service**: ask about this

THE MICROWAVE

The microwave oven produces energy waves called microwaves. These waves hit food and cause the molecules in it to vibrate. This produces heat, which cooks the food. Some microwave ovens are more powerful than others. They range from 600 watt to 1000 watt (1000 watt is the most powerful).

Microwave oven

Advantages and uses of microwave ovens
- cooks food very quickly; suits today's busy lifestyle
- healthy cooking method
  - no extra fat added
  - fast cooking – little loss of nutrients
- thaws food quickly; suits today's busy lifestyle
- can reheat food quickly and thoroughly
- small, convenient appliance
- melts jam, chocolate, jelly etc. quickly but be careful not to overheat
- ripens avocado pears, bananas or tomatoes (on low heat; pierce skin)

Safe use of the microwave
- do not put anything metallic into the microwave
- prick anything with a skin, e.g. tomatoes, sausages etc., before cooking to prevent bursting
- be careful to mix foods thoroughly before eating: food may be cool on the outside and red hot in the middle
- leave to 'stand' for a few minutes before eating as food continues to cook during this time

Buying a microwave
Read the general guidelines above for buying electrical appliances. Also, consider these special features that may be available in a microwave:
- has the oven a turntable? (better for even cooking)
- what is its wattage? (the higher the wattage the more powerful the oven)
- does it have a browning element?
- does it offer dual control? (cook half by microwave half by conventional oven)

Activity 29.3 – Workbook p. 231

Exam time – Workbook p. 231 – The microwave oven

THE COOKER

Types of cooker

Gas cooker

Electric cooker with gas hob

Split level: hob can be built into the worktop and the oven at waist level: one can be gas and the other electric

Solid fuel cookers often heat water and radiators as well as cook but they are very expensive to buy

ESSENTIALS FOR LIVING

Modern features

Fan ovens: a fan oven blows hot air around the oven: cooks food more quickly (set cooking temperature 10°C lower)

Top oven: grill can also be used as a small oven

Ceramic hob

Ceramic hob: easy to clean, heat-resistant glass surface (see photo)

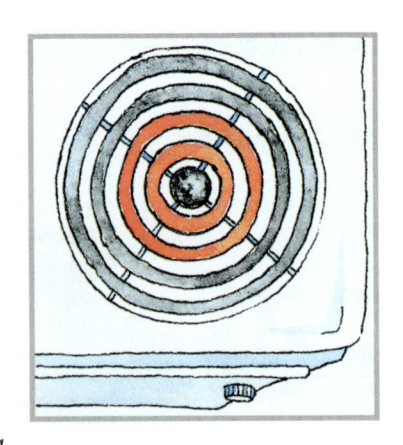

Dual ring

Dual rings: can be set so that only the middle part of the ring heats, for small saucepans (see diagram)

Dual grill: full grill or one side only heats up for small amounts of food

Self-cleaning ovens: door locks; oven reaches very high temperatures; food burns off; ash can then be swept out

Autotimers: three clocks, may be digital (see below)

Timers are less complex. You set the timer for a certain amount of time, for example one hour; the oven switches on there and then (it cannot be set for later in the day, like autotimers). It switches off when the time is up.

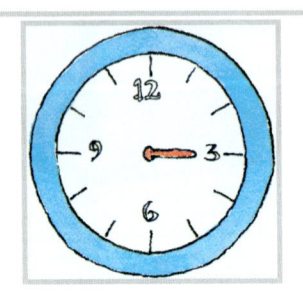

| Set this clock for the time you want the oven to come on at, e.g. 3 p.m. | Set this clock for the time you want the oven to switch off at | This clock should be set at the present time, e.g. 7.30 a.m. |

Autotimer

244

Buying a cooker

1. See general guidelines for buying electrical appliances page 241.
2. Should you buy a gas or electric cooker? If your house does not have natural gas you may wish to choose an electric one as bottled gas can be inconvenient. Gas cookers, on the other hand, heat up quickly and are easily controlled.
3. Consider whether you want or can afford extra features such as a ceramic hob, split-level cooker. What extra features are important to you? For example, an autotimer?

Care and cleaning of cookers

1. Wipe up spills straight away; why?
2. Wash grill pan after each use; why?
3. Never drag saucepans over a ceramic hob, they will scratch it. Use a special ceramic hob cleaner.

4. If using oven cleaners, protect yourself and your surroundings: they are dangerous.

Activity 29.4 – Workbook p. 232

Exam time – Workbook p. 232 – Cookers

THE REFRIGERATOR

Types of refrigerator

Fridge with icebox

Fridge-freezer

Features of a modern refrigerator

- thermostatically controlled
- door of fridge can be made to match kitchen units
- cold drinks dispenser
- moulded door storage and moveable plastic-coated shelves
- automatic defrost: the fridge or fridge-freezer switches itself off every now and then; ice does not build up so the icebox or freezer is kept defrosted
- salad drawers

Choosing a refrigerator

See general guidelines for buying electrical appliances page 241.

Rules for using a refrigerator

1. Do not position the fridge near the cooker or a radiator: its motor will have to work harder to keep the fridge cool.
2. Never put hot food in the fridge.
3. Cover food before putting it in the fridge: this stops it drying out and absorbing strong smells such as onions.
4. Do not over-pack the fridge.
5. Put meat (especially raw meat) on a plate so it does not drip onto other food. Store in the coldest part of the fridge (near icebox).
6. Close the door immediately after use.

Care and cleaning

Defrosting the icebox

1. Choose a time when the icebox or freezer is fairly empty.
2. Some models have an automatic defrost (see above), if not:
   - remove all food and wrap in a good insulator, e.g. newspaper
   - allow ice to melt into the drip tray; empty the drip tray
   - when all the ice is gone wash out and dry the icebox or freezer
   - return the food items
   - never use a knife to remove ice

Cleaning the refrigerator

1. Choose a time when the fridge is fairly empty.
2. Remove all food and wrap in a good insulator, e.g. newspaper.
3. Remove all moveable shelves and parts.
4. Clean the fridge with a solution of bread soda (2 tablespoons) and water (1 litre).
5. Rinse with plain, warm water. Dry. Replace shelves and food.
6. Wash the outside with warm water and washing up liquid. Dry well.

Exam time – Workbook p. 234 – The refrigerator

Now test yourself at *www.my-etest.com*

Chapter 30

Home hygiene

A good standard of hygiene in the home is essential for good health. Bacteria thrive in damp, dirty conditions and cause food poisoning and disease.

Guidelines for good hygiene in the home

1. Ensure good ventilation: bacteria love stuffy, damp conditions
2. Keep the house warm and dry
3. Open curtains and windows daily: sunlight and fresh air help destroy bacteria

In the bathroom, avoid carpets, toilet mats or seat covers; why?

4. Kitchens and bathrooms should have smooth, easy-to-clean surfaces such as tiles
5. Empty kitchen bins daily and large dustbin weekly; wash and disinfect regularly
6. Keep toilets and sinks spotless, disinfect once a week

Cleaning agents

Nowadays there are a huge number of different cleaning agents on the market. Many are poisonous, so:

1. Store only in their original containers: dangerous products will often have a childproof container
2. Keep out of the reach of children (do not store under the sink)
3. Store in a dry place
4. Follow directions carefully
5. Wear gloves and protective clothing while using
6. Rinse away all traces of cleaning agent after use

Cleaning agent	Example	Use
water	–	warm (washing) cold (soaking)
detergents	washing powder, soap, washing up liquid, dishwasher powder	washing clothes and dishes
cream cleaners	Cif	kitchen and bathroom surfaces, cookers
abrasive cleaners	Ajax, Brillo	stubborn stains on scratch resistant surfaces, e.g. metal saucepans (not non-stick)
window cleaners	Windowlene	windows – cleans without streaking
polish	Pledge, Mr Sheen (sprays) solid wax	polishing furniture, floors
metal polishes	Brasso, Silvo, silver cloths	polishing metal – brass and silver
bleaches	Parazone, Domestos	kill germs, remove stains
disinfectants	Dettol, Savlon	kill germs

Note: do not use bleaches and toilet cleaners (e.g. Harpic) together as a poisonous gas is given off when they are mixed.

Activity 30.1 – Workbook p. 237

Exam time – Workbook p. 237

Now test yourself at *www.my-etest.com*

chapter 31

Safety and first aid

ACCIDENTS IN THE HOME

Causes
- carelessness or curiosity (children especially)
- badly designed homes, e.g. hidden steps
- faulty equipment
- incorrect storage of dangerous substances

Activity 31.1 – Workbook p. 239

Accident prevention

Electricity

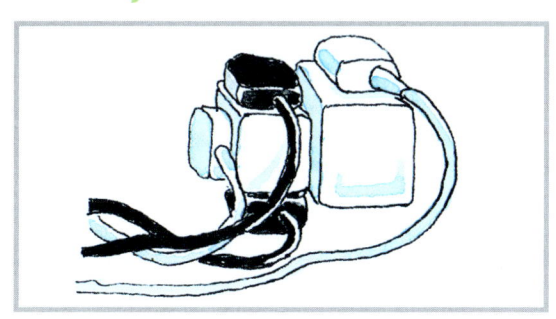

- electricity and water do not mix: never touch anything electrical with wet hands
- wire electrical appliances correctly with the correct size of fuse
- check for safety symbols on appliances
- never take anything electrical such as a heater or hairdryer into the bathroom
- never overload sockets or repair flexes with tape
- fit childproof covers on electric sockets

Fire

- never leave matches or cigarette lighters where children can get them
- fit a full fire guard like the one pictured
- do not hang a mirror over the fireplace; why?
- use an electric deep-fat frier, not a chip pan
- stub cigarettes out completely, never smoke in bed
- check nightwear is fire-resistant (especially children's)
- unplug electrical appliances before going to bed
- close doors at night: stops a fire spreading

249

ESSENTIALS FOR LIVING

Fire safety equipment

Fire blanket: keep in the kitchen; throw over a small fire to smother it

Fire extinguisher

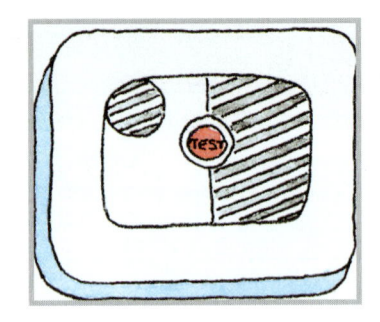

Smoke alarm: check battery on the same day each week

Falls

- have a light switch at the top and the bottom of stairs. Make sure the carpet is not loose. Never leave objects, such as toys, on the stairs
- avoid frayed carpets and over-polished floors
- have grips on the sides of baths and showers. The bottom of the shower or bath should have a non-slip surface (especially for elderly people)
- be careful on newly washed floors; wipe up spills immediately

Children

Accidents are the biggest cause of injury and death of children in this country. Each year approximately one in five children has an accident at home that is serious enough to need treatment by a doctor or in hospital.

All children have accidents, no matter how safe the home or how careful the parents. Nevertheless parents can reduce the risk of accidents by taking the following precautions:

Never leave a young child unsupervised in the house.
Only buy equipment and toys with safety symbols attached.

250

Type of accident	How to prevent it
Choking and suffocation	• keep plastic bags and small hard objects out of reach of children • never leave a baby or a young child alone while feeding • be careful with curtain cords etc.
Scalds	• have a short safety flex on kettles • don't use table cloths; why? • keep hot drinks away from the edges of tables • never drink tea or coffee with a child in your lap • turn saucepan handles inwards • put cold water into the bath first
Falls	• fit window locks • fit stair gates • do not allow children onto balconies alone • strap babies into buggies and high chairs
Poisoning	• keep medicines locked away • medicines should be stored in childproof containers • lock other dangerous substances away also, e.g. alcohol, bleach, weed killer, rat poison
Burns	• fit a fire guard • never leave matches, lighters or petrol within reach • be careful using an iron, don't allow flex to dangle • be sure nightwear is flame-resistant
Drowning	• never leave a child alone in the bath or anywhere there is water

FIRE DRILL

Small fire

1. Use a fire extinguisher or a fire blanket to put out the fire.
2. Never put water on burning oil or an electrical appliance.
3. If a chip pan goes on fire do not bring it outside: the oxygen will feed the fire. Try to put the lid on to smother the flames.

ESSENTIALS FOR LIVING

4. Call the fire brigade if necessary.

Large fire

1. Get out: crawl along the floor: there is more oxygen low down.
2. Call the fire brigade.
3. Do not re-enter the house.

Activity 31.2 – Workbook p. 240

FIRST AID

Aims of first aid

- to preserve life, for example give mouth to mouth resuscitation (see below)
- to help stop the condition getting worse, for example stop the bleeding (see below)
- to promote recovery, for example put in the recovery position (see below)

A serious accident

1. Do not move the patient unless absolutely necessary
2. Check for a pulse (inside of wrist or under ear lobe)
3. If patient is not breathing start mouth-to-mouth resuscitation
4. Don't give the patient anything to eat or drink (they may need to fast for an anaesthetic later)
5. Ring 999: give clear directions to the accident scene

ABC of resuscitation

A is for **AIRWAY**

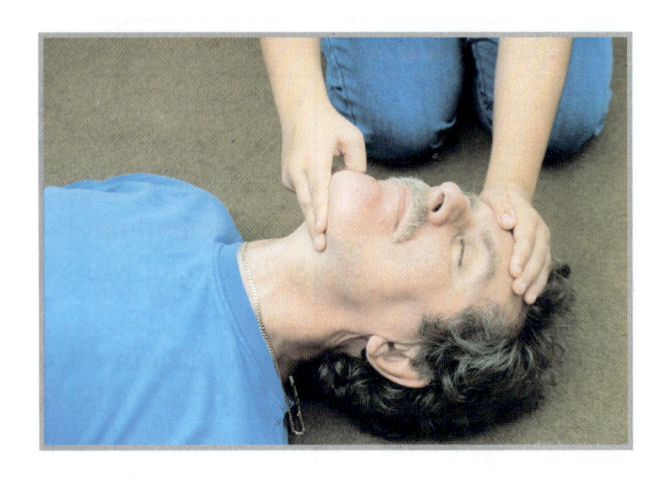

Tilt the casualty's head back and lift the chin. This will open the airway.

B is for **BREATHING**

If the casualty is not breathing, breath for them. Cover their mouth with yours and blow air into their lungs.

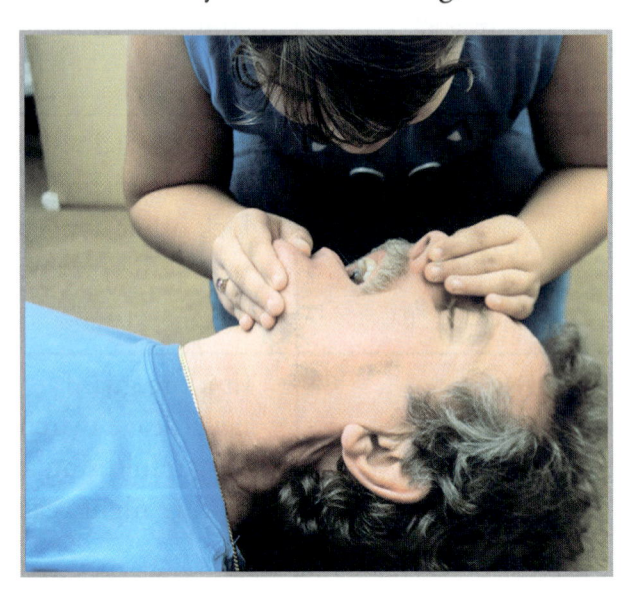

C is for **CIRCULATION**

If the heart stops (no pulse) you can apply chest compressions. Two breaths – fifteen compressions, two breaths – fifteen compressions, and so on.

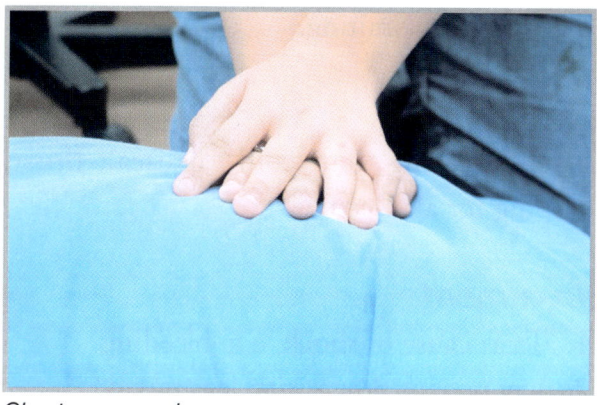

SAFETY AND FIRST AID

Chest compressions

> **Why not do a basic first aid course?**
> Contact The Irish Red Cross society, 16 Merrion Square, Dublin 2. Telephone: (01) 676 5135, email: *redcross@iol.ie*
> They will put you in touch with your local branch.

First aid kit

Every home should have a well-stocked first aid kit.

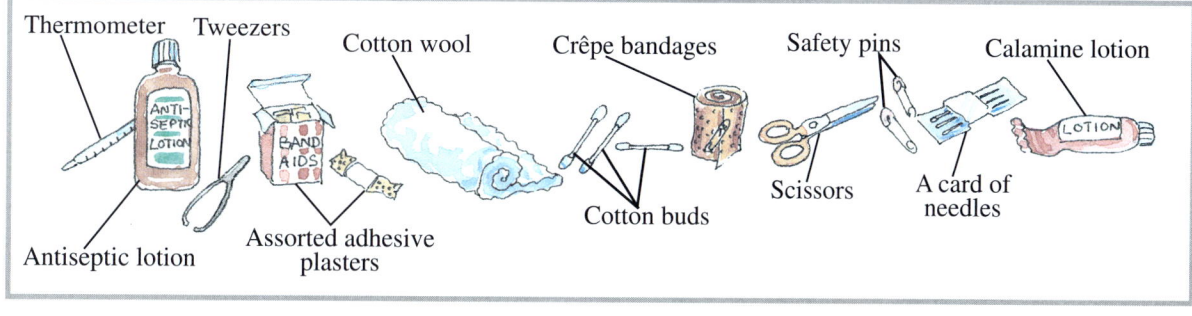

Thermometer, Tweezers, Cotton wool, Crêpe bandages, Safety pins, Calamine lotion, Scissors, A card of needles, Cotton buds, Assorted adhesive plasters, Antiseptic lotion

🌡 Activity 31.3 – Workbook p. 240

BASIC FIRST AID TREATMENTS

Burns and scalds
Minor (not serious)
1. Stop the burning: pour cold water on the burn for ten minutes
2. Gently remove rings, watches, belts etc. before the area swells
3. Reduce the risk of infection: cover with a clean non-fluffy dressing
4. **Do not put any creams on the burn**

Major (serious)
1. Put out the fire by wrapping the victim in a blanket or coat or by rolling them on the ground
2. Call an ambulance
3. Pour cold water on the burn
4. Resuscitate if necessary
5. Don't remove any clothing stuck to the burn
6. Gently remove rings, watches etc. before the area begins to swell

Bleeding

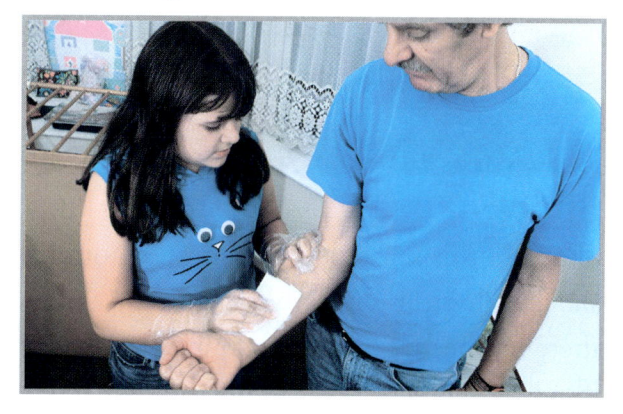

1. Try to stop the bleeding by applying pressure to the wound with your fingers for ten minutes
2. If there is glass or something else in the wound press on either side of it
3. If it is a limb raise it up to help stop the bleeding
4. Wrap in a bandage
5. Call an ambulance

Choking

A child: place face down on your lap with the head lower than body. Slap sharply between the shoulder blades.

An adult: slap sharply between the shoulder blades: if this doesn't work try the Heimlich manoeuvre (see photo).

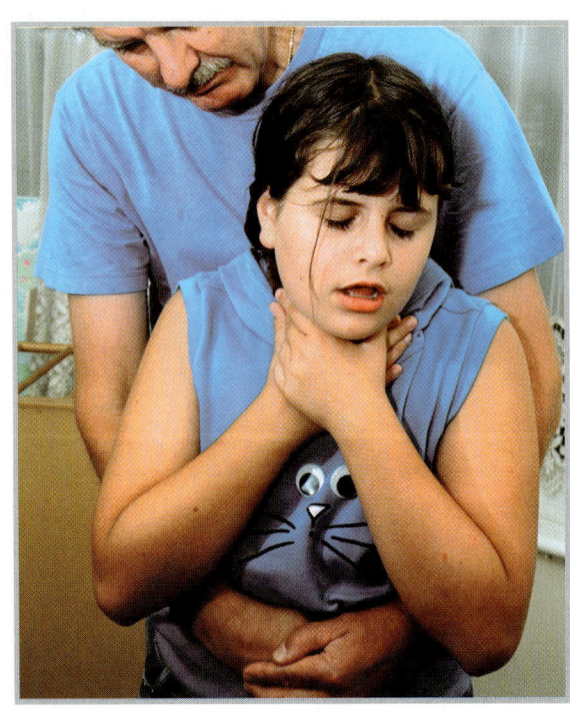

Shock

Shock in this sense does not mean emotional shock. Physical shock usually occurs when there has been a heart attack, a severe loss of blood or a loss of body fluids through severe diarrhoea or vomiting.

Symptoms
Pale, sweating but cold, nausea, thirst, gasping for air

Treatment
1. Treat cause of shock, e.g. bleeding
2. Lay patient down with legs raised
3. Loosen clothing; cover with a coat or blanket
4. Don't give anything to eat or drink

Poisoning

Approximately 170 people die each year in Ireland from poisoning.

1. Call an ambulance
2. Find out what patient has taken, bring the container or the patient's vomit to the hospital
3. Do not try to make the patient vomit
4. Resuscitate if necessary: put in the recovery position

The recovery position is the safest position for the casualty

Exam time – Workbook p. 241

Now test yourself at *www.my-etest.com*

chapter 32

Community services and the environment

ENVIRONMENTAL POLLUTION

Environmental pollution is the term used to describe the ways in which the waste products of human activity harm the natural environment.

Waste products can be:
- Organic or biodegradable: this means that they can be easily broken down by nature, e.g. paper, vegetable peels etc.
- Inorganic: this means that the waste cannot be easily broken down and is harmful to the environment, e.g. plastic, glass and metal.

Activity 32.1 – Workbook p. 244

Environmental pollution is one of the most serious problems facing mankind and all other animals and plants on our planet today. Most people, when asked, say 'yes' they would like to see pollution reduced. Unfortunately, most of the pollution that threatens our health and the health of the planet comes from products people want and are unwilling or unable to do without. For example cars create a large percentage of the world's air pollution.

IMPORTANT ENVIRONMENTAL ISSUES

Air pollution

According to the World Health Organisation about one-fifth of the world's people are exposed to dangerous levels of air pollution. Air pollution occurs when factories and vehicles etc. release large amounts of gas, smoke and dirt into the air. The burning of rubbish in incineration plants can release smoke and heavy metals such as mercury into the atmosphere.

Effects of air pollution

Air pollution causes:
- lung problems such as asthma, bronchitis and cancer
- acid rain (see below)
- the greenhouse effect (see below)

ESSENTIALS FOR LIVING

Smog

Smog is one of the most common types of air pollution. Smog is a brown, hazy mixture of gases and smoke. Smog forms when these gases react with sunlight; this reaction creates hundreds of harmful chemicals which make up smog. Smog is still a problem in many cities today. Dublin had a smog problem in the 1980s, which prompted the government to allow only smokeless fuels to be burned in the city.

Ozone layer

The ozone layer is a layer of gas (O_3) found nine to eighteen miles above the earth's surface. Ozone is very important in that it shields us from the sun's harmful (cancer-causing) UV rays. During the 1970s scientists discovered that a hole was forming in the ozone layer. The main reason for this was found to be substances called CFCs. CFCs were found in aerosol sprays, fridges, fast food cartons and air conditioning systems. Nowadays there is a ban on CFCs in most countries in the world, and the ozone layer is now repairing itself.

Greenhouse effect

Flooding: an effect of global warming

Burning fuel (cars, factories, home heating) produces greenhouse gases such as carbon dioxide. These gases trap heat on the earth's surface, raising the temperature (global warming). The main effect of global warming, if it continues, will be flooding of coastal areas.

Water pollution

Water pollution is the contamination of water by sewage, chemicals, metals, oil and other substances.

COMMUNITY SERVICES AND THE ENVIRONMENT

According to the World Heath Organisation approximately five million people die every year from drinking polluted water – that is more than the population of Ireland. The main effects of water pollution on humans are diseases such as cholera, dysentery and E. coli poisoning (see page 39). Water pollution also causes the death of fish and damage to birds, plants and other wildlife.

Waste disposal

One of the biggest problems facing society is waste disposal. Waste can be disposed of in two ways: it can be dumped and buried in landfill sites or burned in incinerators. Both of these methods can be damaging to us and the environment. We need to produce less waste.

Activity 32.2 – Workbook p. 244

PROTECTING THE ENVIRONMENT

What can you do to help protect the environment?

1. Be an eco-friendly shopper:
   - avoid over-packaged goods
   - look for the EU ECO label (this shows that the product is environmentally friendly and does not harm the environment)

ECO friendly label

2. Recycle:
   - bottles, cans, clothes, paper; there is probably a recycling bank near you – it just takes a little effort

Recycling symbol

   - recycle clothes by donating them to charity shops
   - recycle organic waste, such as vegetable peels, by having a compost bin or compost heap (some local authorities are supplying compost bins at a reduced price)

3. Don't litter and don't tolerate those who do

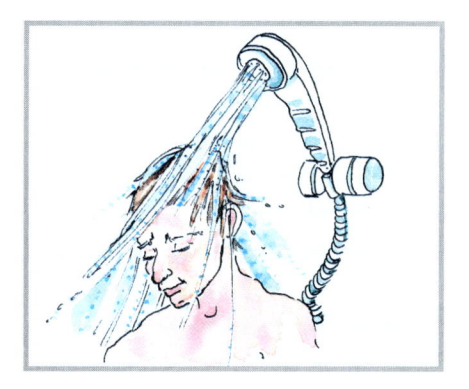

4. Use less water:
   - showers use less than half the water baths do

ESSENTIALS FOR LIVING

- fix dripping taps
- don't turn on the dishwasher or washing machine for small amounts. Use half-load or economy wash if you must wash less than a full load or if the load is not very dirty
5. Walk or cycle to school where possible
6. Use less energy:
   - do not over-heat the home
   - turn off lights when not in use
   - use energy saving bulbs (CFLs)
7. When choosing electrical appliances, choose energy-efficient models: A or B labels (see page 241)

Why not visit our government-sponsored interactive website to find out more? *www.10steps.ie*

The plastic bag levy (15c per plastic bag)

This levy was introduced on 4 March 2002 to encourage consumers to cut down on the number of plastic bags they use. Do you think it is working?

What other measures do you think the government could introduce to help us help the environment?

Organisations involved with environmental protection

- Greenpeace
- Friends of the Earth
- The Green Party
- Environment Protection Agency

Exam time – Workbook p. 245 – The environment

COMMUNITY SERVICES

When people live together in urban (city or town) or rural (country) areas they form a *community*.

Most communities have a number of services and amenities.

Services

Statutory (government-run)
- health (hospitals, health centres)
- education (schools, some pre-schools)
- social welfare
- libraries
- local authority housing (council or corporation)
- Gardaí
- public transport

Voluntary services
- Homeless Aid
- ISPCC
- Society of St Vincent de Paul
- REHAB Foundation
- GAA
- youth clubs
- Samaritans

COMMUNITY SERVICES AND THE ENVIRONMENT | 32

GAA – voluntary service

Amenities

An *amenity* is another word for a leisure facility. Amenities can be *natural*: rivers, lakes, beaches etc., or *man-made*: parks, playgrounds, sports pitches, cinemas, museums, fitness centres, swimming pools etc.

Activity 32.3 – Workbook p. 247

Exam time – Workbook
p. 248 – Community services

Now test yourself at *www.my-etest.com*

Unit 5

Textile Studies

chapter 33

Textiles

What are textiles?

A textile is a fabric or cloth. The clothes we wear and many of the items in our homes are made from textiles.

Properties of textiles

Textiles can have many different properties. They can be:

- cool
- soft
- warm
- absorbent
- waterproof
- washable
- colour-fast (won't run)
- soft
- shiny
- rough
- smooth
- delicate
- light
- heavy
- hard-wearing
- breathable
- non-absorbent
- crease-resistant
- flame-resistant
- resilient (bounces back into shape)
- stretchy
- stain-resistant.

When choosing a fabric for a particular purpose, for example a football jersey, the properties of the fabric you choose are important. Imagine buying a football jersey made from a heavy, non-absorbent, dry-clean-only fabric.

Activity 33.1 – Workbook p. 251

One of the most important uses of textiles is clothing. Natural fabrics such as cotton, wool and linen are used together with man-made fabrics such as nylon, polyester and acrylic.

Functions of clothing

There are six basic functions of clothing:

1. Protection from the weather

TEXTILES 33

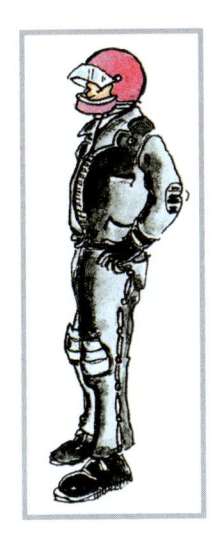

2. Safety

3. Modesty: levels of modesty vary between cultures

4. Self-expression

5. Identification

6. Creating an image

✏️ **Activity 33.2 – Workbook p. 251**

TEXTILES IN THE HOME

Textiles have many uses in the home
- curtains
- cushions
- bed linen
- towels
- oven gloves

ESSENTIALS FOR LIVING

- lamp shades
- carpets
- sofas
- armchairs
- blinds
- pillows
- wall hangings

Textiles in the home

Choosing fabrics for household items

Points to consider:
- cost: some fabrics, for example silk, are very expensive
- durability: fabrics need to be hard-wearing
- washable: fabrics ideally should be washable
- colour and pattern: should fit in with the décor already there

Curtains

Functions
- provide privacy
- keep draughts out
- keep heat in
- decorate a room

Curtains insulate and decorate

Properties
Curtains should:
- hang well
- be fade-resistant
- be washable (ideally)
- be fire-resistant
- be pre-shrunk
- have a close weave

Measuring curtains
Length: from rail to 5 cm below window sill (short curtains); from rail to 2 cm off the floor (long curtains)
Width: 2.5 times width of window or rail
Suitable fabrics: cotton, linen, velvet, dralon (best lined)

Carpets

Carpets can be one of the most expensive purchases for the home. Carpets can be:
a) woven (expensive)
b) bonded (less expensive)

Properties

A good carpet will be:

- resilient (bounce back after being walked upon)
- warm
- hard-wearing
- stain-resistant

What type of carpet you buy will depend on which room it is for and how much money you have to spend. Ideally a room with a lot of traffic, such as a living-room or a hall, should have a quality carpet, for example a woven eighty per cent wool/twenty per cent nylon mix. A less expensive carpet, on the other hand, would be fine for a bedroom.

Classification of carpets

Most carpets carry a label suggesting where it is suitable for:

Light domestic: bedrooms
General domestic: living-rooms
Heavy domestic: halls and living-rooms

Other household textiles

Household item	Made from	Points to remember
bed linen • sheets • pillow cases • duvet covers	cotton, polycotton, polyester	needs to be washable, easily dried and absorbent
duvets	cover: cotton, polyester or polycotton; filling: polyester or duck down	warmth of a duvet is measured by a tog rating 4.5–15 tog (warmest) duvets need to be warm, light and washable
upholstery • covers on sofas, chairs • mattresses: cover and filling	cotton, dralon, polycotton	needs to be durable, stain-resistant, closely woven and fire-resistant

Exam time – Workbook p. 252

Now test yourself at *www.my-etest.com*

chapter 34

Fashion and design

Fashion is a word used to describe clothes or other items that are trendy or popular at any given time. Fashion changes constantly. *Fashion trends* are changes in fashion. Most people also see that fashion trends form cycles of fashion, for example 1970s fashion (flares) returned in an almost identical form in recent years.

Fashion 2002

Activity 34.1 – Workbook p. 254

Fashion trends are influenced by:
1. The fashion industry
2. Famous people
3. Historical events and the economy
4. Technology

1. The fashion industry

Twice every year fashion shows are held in Paris, London, New York and Milan. Here the world's top fashion designers show off their designs. These *haute couture* clothes are hand-made, one of a kind and *very* expensive (€20,000 for a suit).

Fashion 1972

FASHION AND DESIGN 34

Haute couture – high fashion

Top designers such as Prada, Versace, and Calvin Klein also produce less expensive machine-made clothes. These ranges, called *prêt-à-porter* (ready to wear) clothes, are still expensive and only available in designer boutiques.

Chain stores such as Sasha, NEXT, Cocoon, and Dunnes Stores then buy or copy these designs and sell them cheaply as off-the-peg clothes.

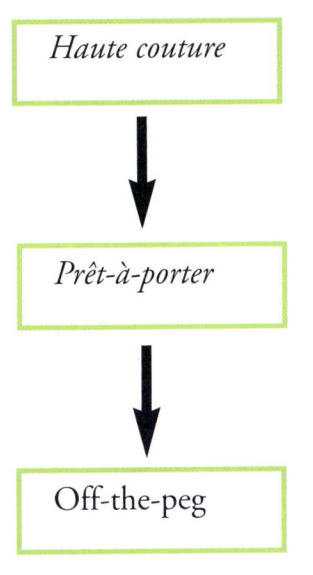

Haute couture
↓
Prêt-à-porter
↓
Off-the-peg

2. Famous people

Pop stars, sports stars, models, actors and actresses all influence current fashion trends through the media.

Famous people influence fashion trends

267

3. Historical events and the economy

- during World War II fabric was scarce, so skirts became straight and shorter
- during times of economic depression in a country, aggressive, anti-establishment fashions become more popular; for example punk, goth etc. Ironically, nowadays this type of fashion is more common among middle-class teenagers

Street fashion

4. Technology

As new fabrics are developed they influence fashion.

Activity 34.2 – Workbook p. 254

BUYING CLOTHES

What influences your choice?

Age: as you get older your idea of what is nice changes.

Cost

Fashion: often fashion influences our likes and dislikes. For example when flared jeans are in fashion most people dislike drainpipes (straight, tight leg).

What suits you

Peer pressure

Lifestyle and occupation: a student would be likely to dress differently to a bank manager for example.

Culture: some cultures, for example Muslims, have a dress code.

Factors or guidelines you should consider when buying an item of clothing:

- cost: can I afford it?
- need: do I need it?
- does it suit my figure, hair colour, skin colour?
- fit: does it fit me well, is it comfortable?
- quality: is it good value for money?
- is it washable?
- does it suit the occasion, e.g. a job interview?

Accessories are worn with an outfit to complete it. Examples include jewellery, belts, hats, shoes, bags etc. Some accessories also have a function, for example a bag is used to carry things.

FASHION AND DESIGN **34**

Exam time – Workbook p. 254

DESIGN USING TEXTILES

When designing clothes or household items with textiles there are two basic aims.
The item must:
   a) function as it should
   b) look well

As with interior design (see page 224) certain design guidelines or factors, when followed, can achieve these aims.

Design guidelines or factors

1. **Function**: the properties of the fabric you choose should allow the finished product to function as it should. What properties would be important in fabric for an apron?
2. **Colour**: choose colours that suit your hair, skin and eye colour; for example bright red may not suit a person with red hair. Some colours never go out of fashion; for example black, white, navy and denim.
3. **Pattern**: avoid too much, especially large patterns.
4. **Texture**: the feel of a fabric. Can you name a clothing fabric with a smooth or a rough texture?
5. **Line**: horizontal lines (stripes) make a person look wider. Vertical lines make a person look taller and slimmer. Curved lines, such as those in a flowing dress, are feminine.
6. **Shape**: the shape or outline of a garment. A well-cut garment can flatter any figure.

The same principles of proportion, emphasis, balance and rhythm also apply to clothing and textile design (see interior design page 224).

The outfit pictured here will function well as a work suit; it will not date as black and white are colours that do not go out of fashion. While there is quite a lot of pattern, it works because colours remain the same. The suit has a well-cut, rectangular shape which flatters the figure.

Activity 34.3 – Workbook p. 256

Now test yourself at *www.my-etest.com*

chapter 35

Fibres and fabrics

Fabrics are made from fibres. Fibres can be divided into two groups: natural fibres and man-made fibres. Each group can be divided again.

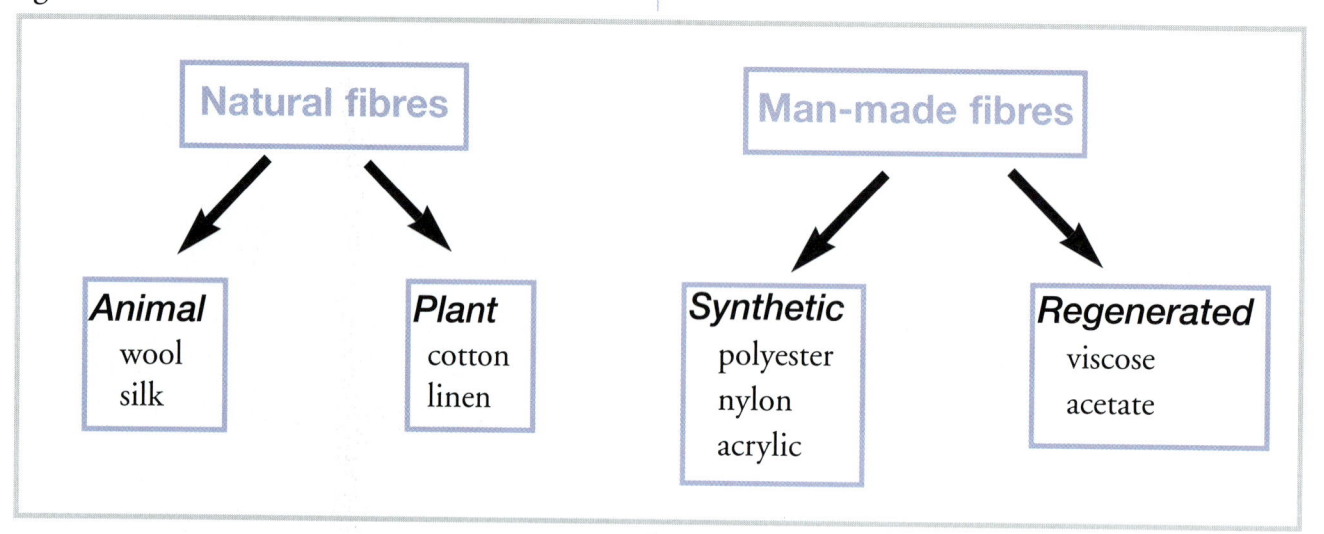

Natural fibres
- *Animal*: wool, silk
- *Plant*: cotton, linen

Man-made fibres
- *Synthetic*: polyester, nylon, acrylic
- *Regenerated*: viscose, acetate

NATURAL FIBRES (ANIMAL)

Wool

Source and production

1. Sheep are sheared (also goats, camels, rabbits)
2. Fleece is graded, washed and dried
3. Wool is brushed (carded) and combed
4. Strands are spun into yarn: long strands (worsted) and short strands

FIBRES AND FABRICS — 35

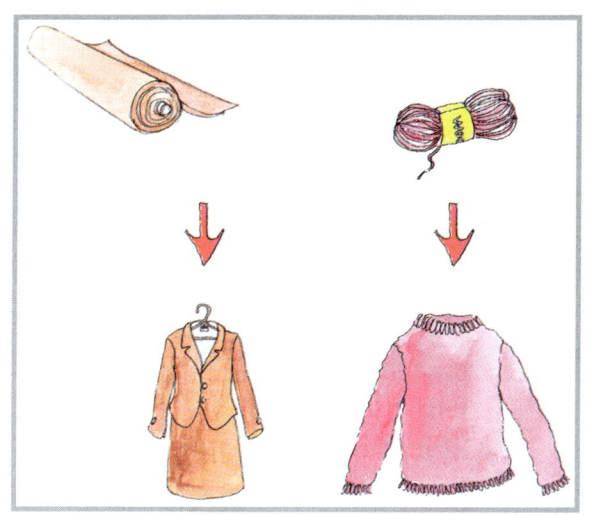

Long strands (worsted) are used to make wool fabric

Short strands are used for knitting wool

100% wool   Wool blend (wool + another fibre)

Properties

☺ Good: wool is a warm, absorbent fabric that does not burn easily. Wool carpets are hard-wearing and resilient (don't stay flattened when walked on).

☹ Bad: wool does, however, shrink if washed at too high a temperature or if tumble dried. It can also irritate the skin.

A suit made from wool

Uses of wool (short fibres): jumpers, blankets, carpets

Uses of wool (long fibres): long worsted fibres are used to make different types of wool fabric; examples include gabardine, tweed and crepe

✏️ **Activity 35.1 – Workbook p. 257**

ESSENTIALS FOR LIVING

Silk

Source and production

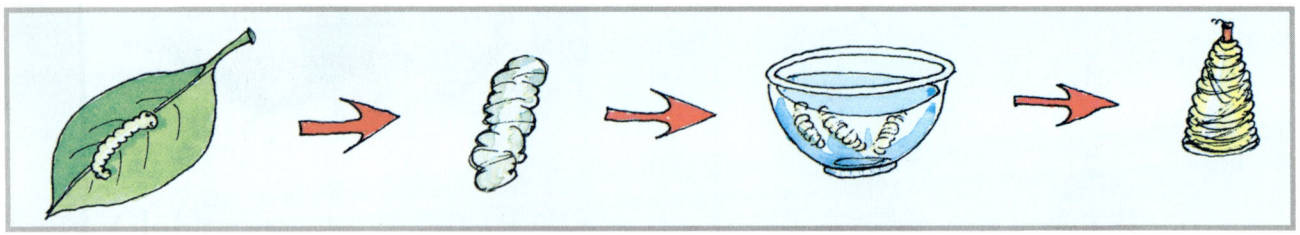

1. Silk worm feeds on mulberry leaves
2. Silk worm spins a cocoon of silk around itself
3. Cocoons are soaked in water
4. Silk threads are unwound from cocoons onto reels
5. Several of these thin silk threads are spun or twisted together to make thicker thread; these are then woven into fabric

Silk symbol

Properties
☺ Good: silk is absorbent, crease-resistant, strong, smooth, light and drapes (falls) well.
☹ Bad: silk is expensive, also easily damaged by careless washing, moths, sunshine and chemicals. It is also flammable (burns easily).

Uses of silk
Silk is a luxury fabric and is used for shirts, evening dresses etc. In the home silk can be used for paintings, cushion covers, curtains (line them; why?).

Types of silk
Chiffon (very light), taffeta, wild silk, satin

Silk evening dresses

FIBRES AND FABRICS

NATURAL FIBRES (PLANT)

Cotton

Source and production

Cotton bolls

↓

picked
cleaned
graded
combed
spun

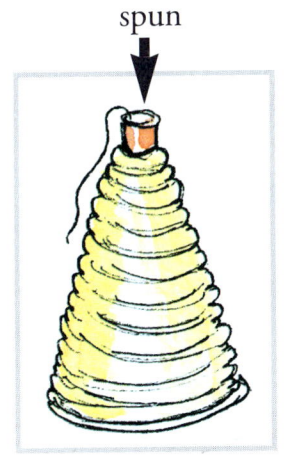

↓

Yarn

100% cotton

Properties

☺ Good: cotton is absorbent, cool, strong, washes and dries well and is easy to dye and bleach.

☹ Bad: cotton creases, shrinks and burns easily. It can be damaged by mildew (mould).

Denim is one of the most popular types of cotton

ESSENTIALS FOR LIVING

Uses of cotton
Clothes, towels, sheets and curtains

Types of cotton
- flannelette (fluffy sheets)
- denim
- towelling (towels)
- muslin (light see-through fabric)

Linen

Source and production

1. Flax plants are grown in wet countries like Ireland
2. Flax stems are soaked until the woody part rots (retting)
3. Fibres are separated from the woody stem. They are then bleached and spun into yarn

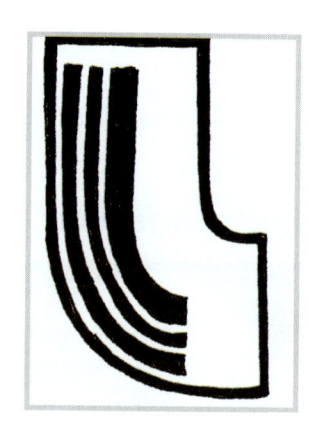

Linen

Properties
☺ Good: linen is absorbent, cool, strong and washes well.
☹ Bad: linen is expensive, creases and shrinks easily. It is difficult to dye and is damaged by mildew.

Irish linen products

Uses and types of linen
Irish linen is famous the world over. Linen products range from heavy damask tablecloths and sheets, to finer lawn and cambric clothing.

✏ Activity 35.2 – Workbook p. 257

Revision crossword – Workbook p. 258

MAN-MADE FIBRES

Man-made fibres can be either *regenerated* or *synthetic*.

Regenerated

Viscose, rayon, acetate

Sources and production

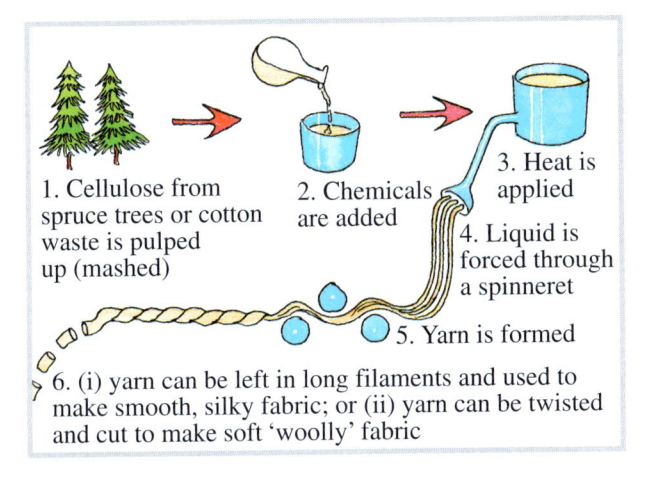

1. Cellulose from spruce trees or cotton waste is pulped up (mashed)
2. Chemicals are added
3. Heat is applied
4. Liquid is forced through a spinneret
5. Yarn is formed
6. (i) yarn can be left in long filaments and used to make smooth, silky fabric; or (ii) yarn can be twisted and cut to make soft 'woolly' fabric

Uses

Fabrics made from regenerated fibres such as rayon and viscose are used to make silky blouses, dresses etc. (see photograph)

Properties

☺ Good: regenerated fabrics are absorbent and cool to wear.

☹ Bad: they are not hard-wearing and tear easily after a lot of washing.

ESSENTIALS FOR LIVING

Synthetic
Nylon, polyester, acrylic

Sources and production

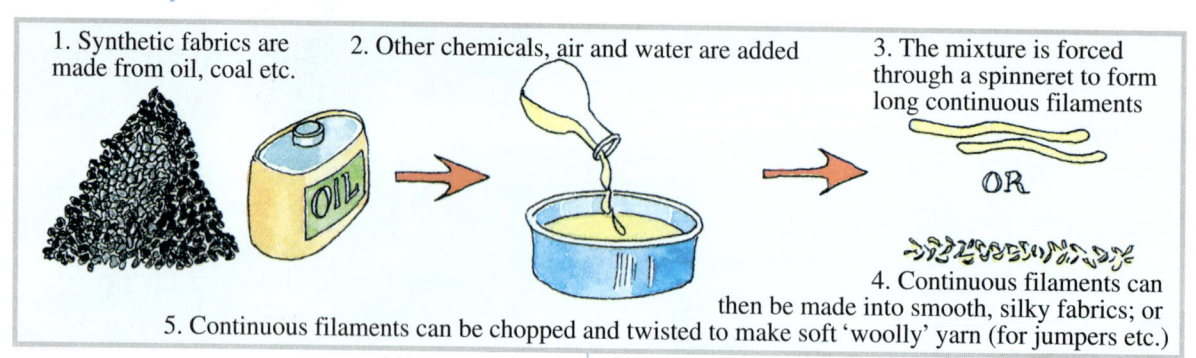

1. Synthetic fabrics are made from oil, coal etc.
2. Other chemicals, air and water are added
3. The mixture is forced through a spinneret to form long continuous filaments
4. Continuous filaments can then be made into smooth, silky fabrics; or
5. Continuous filaments can be chopped and twisted to make soft 'woolly' yarn (for jumpers etc.)

Acrylic jumper

Uses of synthetic fibres
Acrylic: jumpers, fake fur, dresses, carpets
Nylon: umbrellas, swimming suits, tents, raincoats, tights, blended with wool for carpets
Polyester: usually blended with cotton in trousers, blouses, shirts; used to make the filling in duvets and coats

Properties
☺ Good: synthetic fabrics are all strong and hard-wearing. Acrylic and polyester are warm (acrylic jumpers, polyester filling in duvets and coats). Nylon repels water: good for tents, rain gear etc.
☹ Bad: all synthetics are flammable. Some, such as nylon, cling and are clammy in summer as they don't absorb moisture.

✏️ Activity 35.3 – Workbook p. 259

✏️ Activity 35.4 – Workbook p. 259

MAKING FIBRES INTO FABRICS

Yarn
Fibres such as cotton, linen, polyester and wool are twisted into *yarn*. Yarn can be (a) straight (makes smooth fabric), or (b) crimped, looped or short fibres twisted together (makes textured fabrics) – see diagram. Yarn is then treated in one of two ways to make fabric. (*Bonded* fabrics are made from fibres that have not been spun into yarn.)

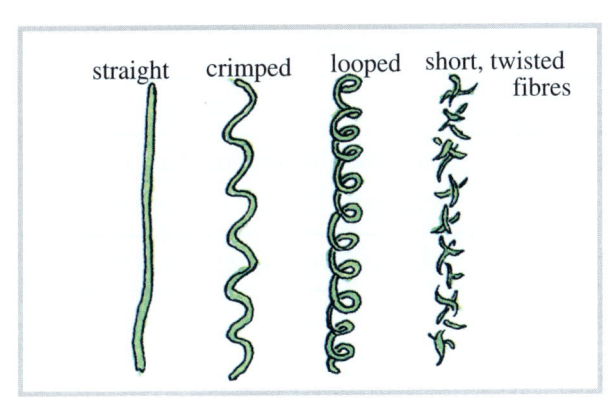

Types of yarn

Yarn to fabric

Weaving

Most weaving is done on a loom. Warp threads are put onto the loom first and weft threads are then woven through them. The selvedge is the edge that the weft threads make; it doesn't fray.

Knitting

Knitting can be done by hand or by machine. Most people, when they think of knitting, think only of jumpers, hats and scarves. Many fabrics are also knitted as you can see if you look closely at them. All knitted fabrics are stretchy. Tracksuits, T-shirts, jumpers and tights are all made from knitted fabrics.

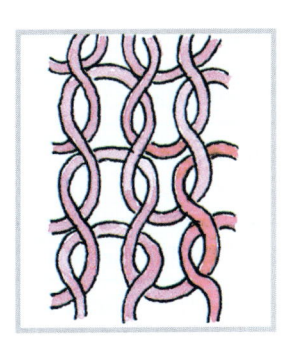

Knitted fabric close up

Bonded fabrics

Bonded fabrics are made from fibres that have *not* been spun into yarn. They are made by sticking fibres together, by applying moisture, pressure and heat.

Examples: Felt (snooker tables, hats), wadding (filling inside duvets, coats, carpet underlay), disposable fabrics (J-cloths, hospital gowns).

The main advantage of bonded fabrics is that they are cheap to make and do not fray. Some, however, are not very long-lasting.

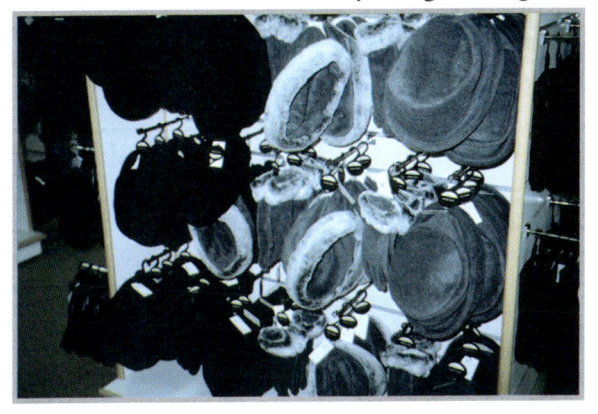

Fabric finishes

A fabric finish is a way of treating a fabric to improve its properties. For example cotton may be treated so that it no longer creases easily (crease-resistant finish).

Activity 35.5 – Workbook p. 260

Fabric finishes

Finish (trade name)	Function	Use
water-proofing (Scotch Guard)	stops water soaking through	anoraks, waterproof sportswear, e.g. golf jackets etc., tents
water-repellent (Scotch Guard)	makes fabric showerproof	showerproof coats, jackets
stain-repellent (Scotch Guard)	stops stains soaking into fabric	clothes, carpets and upholstery
flame-resistant	fabric won't go on fire easily	children's nightwear, upholstery fabrics, e.g. on sofas
brushing	makes fabric feel warmer and softer	flannelette sheets
crease-resistant	reduces creasing, no need to iron	shirts, trousers, sheets, curtains
permanent pleating	pleats won't fall out	skirts and trousers
stretch (Lycra)	woven into fabric to give stretch	swimwear, trousers, cycle shorts, underwear
non-shrink	stops shrinking	clothing, furnishing fabrics
moth-proofing	stops moths attacking fabrics and making holes	wool and silk
anti-static	stops clothes clinging	synthetic underwear, clothing and carpets
polishing (mercerising)	makes fabric smooth and strong	cotton thread and fabric
stiffening	gives a sharp, crisp finish	cuffs and collars

Adding colour and pattern to fabric

Weaving: different coloured yarns are woven into fabric on the loom to make, for example, tartan and checks.

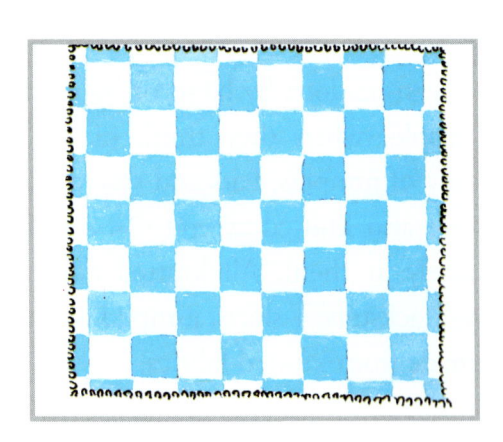

Dyeing

In the past natural products, such as beetroot, were used to dye fabric. Today chemical dyes are used. Chemical dyes, unlike natural dyes, do not run; they are colour-fast. Home dyeing using products such as Dylon can be very effective.

Printing

Printing applies colour to one side of the fabric using a block, screen or roller.

Fibre and fabric identification

Burning tests are used to identify fibres and fabrics. Observe (a) how the test fibres or fabrics burn, (b) the odour given off, and (c) the residue (what is left after burning). (See table below.)

Activity 35.6 – Workbook p. 260

Revision crossword – Workbook p. 261

Exam time – Workbook p. 263

Now test yourself at *www.my-etest.com*

Fibre and fabric identification – burning tests

Fibre	Odour	How it burns	Residue
wool or silk	burning hair	slowly	dark ash
cotton, linen, viscose	burning paper	quickly	paper-like ash
nylon	celery	melts	hard bead

chapter 36

Textile skills

Hand sewing guidelines

1. Use a single thread, not too long
2. Pin and tack if necessary
3. Start and finish with a secure stitch
4. Small, even stitches are strongest
5. A thimble can be used for tough fabrics such as denim

Scissors, measuring tape, thread, pins, needles, stitch ripper, thimble, pinking shears (not essential)

COMMON STITCHES

The best way to learn how to hand sew is:
- ask for a demonstration
- study some samples of the stitches you wish to learn
- practice

Tacking

Uses:

Tacking is a *temporary* stitch. It is used to:
- hold pieces of fabric together for permanent stitching
- help keep machine stitching straight
- hold garment together for fitting

Finish tacking with a backstitch, that is, two or three stitches on top of each other.

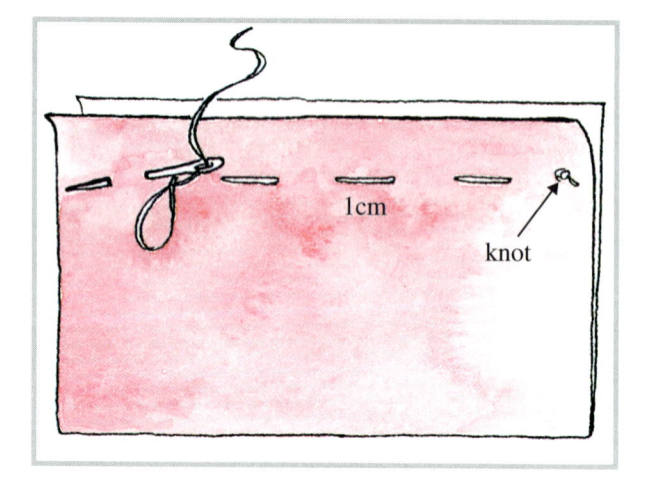

Tacking

Running

Running is like small tacking. Begin with a backstitch.

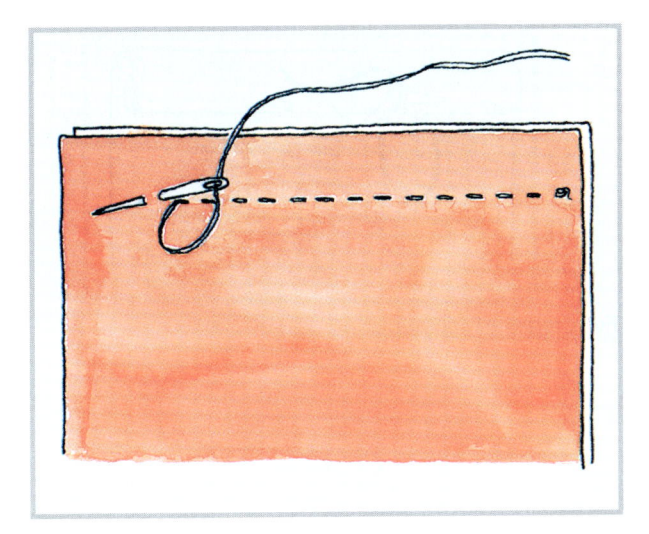

TEXTILE SKILLS 36

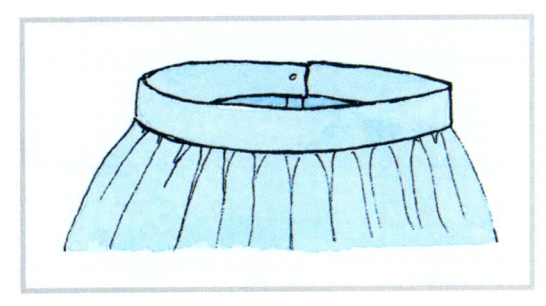

Gathered waistband

Backstitching

Uses:

Backstitching is a strong stitch used for seams instead of machining. Start with a backstitch. Put the needle in at the end of the last stitch and out 2 mm in front. Repeat. There is little or no gap between stitches.

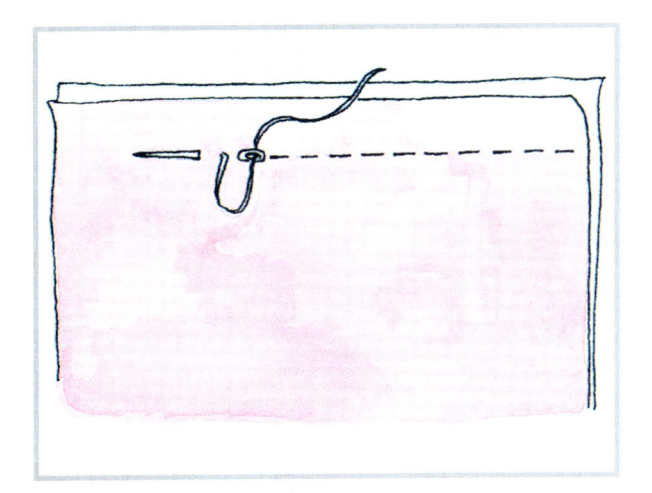

Backstitching

Hemming

Uses:

Hemming is a strong slanted stitch used for stitching down collars and waistbands.

Uses:

Two rows of running stitch are used to gather fabric if you have no machine.

Gathering is used when you wish to join a wide piece of fabric to a narrower one (e.g. fabric into waistband of a skirt).

Gathering (machine or running)

1. Loosen machine tension
2. Machine or run two rows 1 cm from the top of the fabric to be gathered; leave threads
3. Gently pull the gathering threads until you get the correct size; secure with a pin (see diagram)

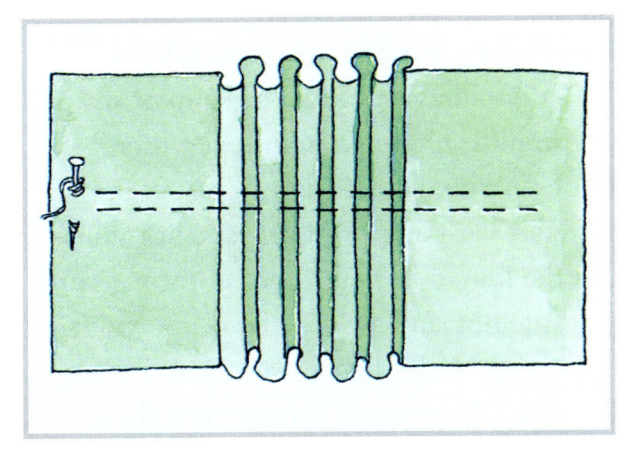

Gathering

281

ESSENTIALS FOR LIVING

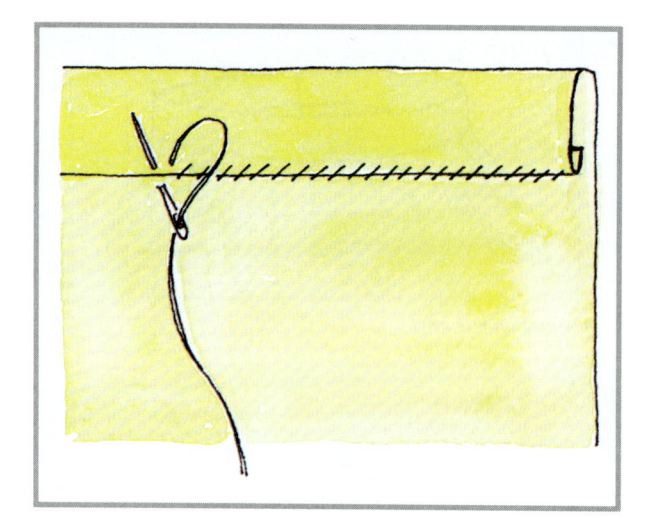

Hemming

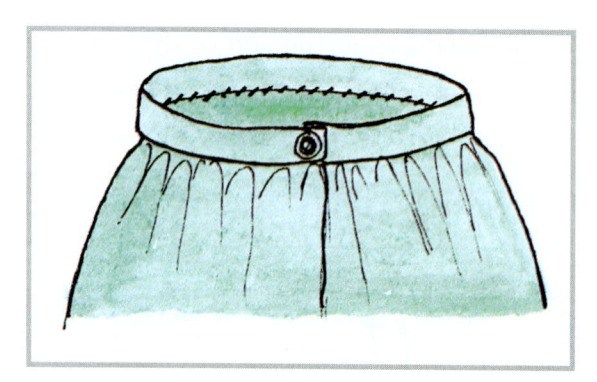

Use of hemming

Slip hemming

Uses:

Slip hemming is used to sew up hems on skirts, trousers and dresses. If well done it is almost invisible on the right side.

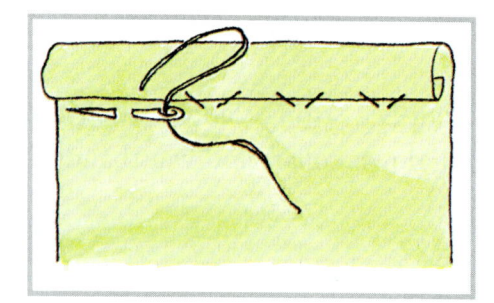

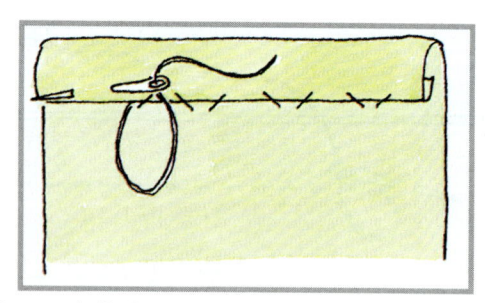

Two stages of slip hemming

Tailor tacking

Uses:

Tailor tacking is used to transfer markings from paper patterns to fabric.

Tailor tacking

EMBROIDERY

Embroidery is used to decorate fabrics.
- embroidery thread is made up of six strands; use three strands at a time
- use *crewel* needles (large eye)
- start with a few running stitches along the line to be embroidered; these secure stitching and are covered as you work embroidery stitches over them
- finish by weaving thread through stitches at the back

TEXTILE SKILLS | **36**

Hand embroidery

Common embroidery stitches

Again, the best way to learn embroidery stitches is to:
- ask for a demonstration
- examine samples
- practice

Skein of thread

Stem stitch
Use: outline stems etc.

Stem stitch

Satin stitch
Use: to fill in small areas such as petals and leaves.

Satin stitch

Long and short stitch
Use: to fill in larger areas.

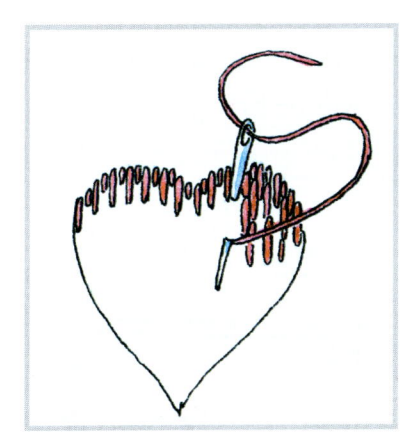

Long and short stitch

Chain stitch
Use: usually to outline.

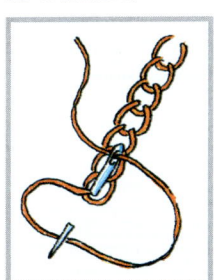

Chain stitch

ESSENTIALS FOR LIVING

Exam time – Workbook p. 268 – Hand sewing

THE SEWING MACHINE

Threading
This varies from machine to machine (read user instructions).
Generally:
- have thread coming from the back of the spool to thread guide (1)
- around tension wheel (2) to take-up lever (3)
- through thread guides (4) and through needle usually from the front to back although can be from left to right depending on the machine
- thread the bobbin and insert; bobbin thread will be brought up by turning the hand-wheel

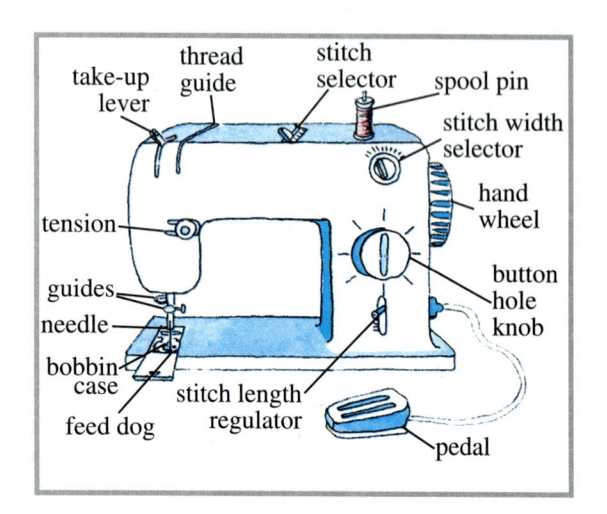

Bobbin and bobbin case

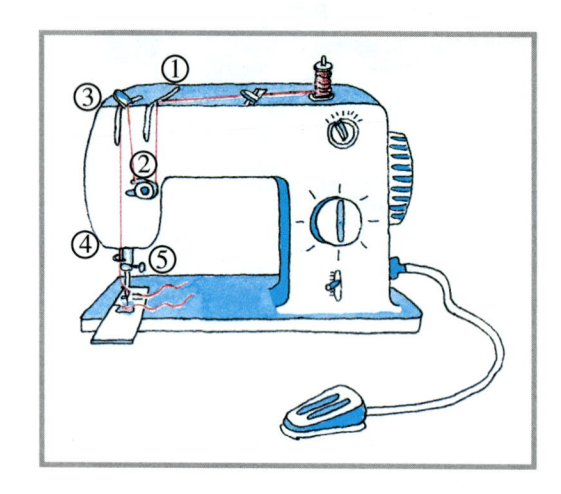

Activity 36.1 – Workbook p. 269

Stitch tension
Two threads form a machine stitch: one from the top spool and the other from the bobbin. Tension (the tightness of the thread) must be equal between both threads. The tension can be adjusted by twisting the tension wheel or regulator on the machine.

TEXTILE SKILLS 36

Correct tension

Top tension too loose

Top tension too tight

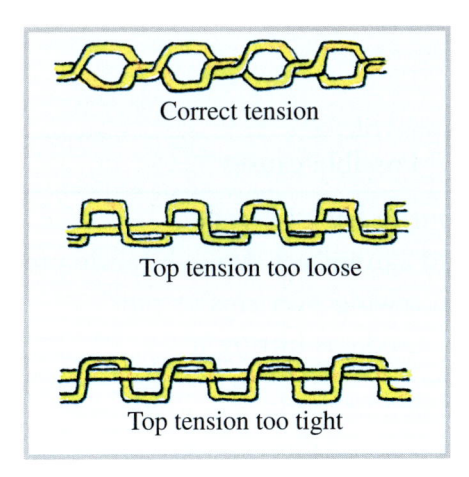

Using the sewing machine

1. Thread the machine correctly using the same thread in both the bobbin and the top spool. Make sure the thread is not too thick for the fabric. Bring both threads away from you towards the back of the machine (if needle threads from front to back).
2. Raise the presser foot and needle, slide the fabric into position and then lower the presser foot and needle into the fabric. Begin machining.
3. Do not push or pull the fabric, just gently guide it.
4. If you must stop in the middle of a line or to turn a corner lower the needle into the fabric (you can then start again exactly where you left off).
5. When you finish you can secure stitching by reversing back a few stitches or by hand sewing a few backstitches.

Common machine stitches

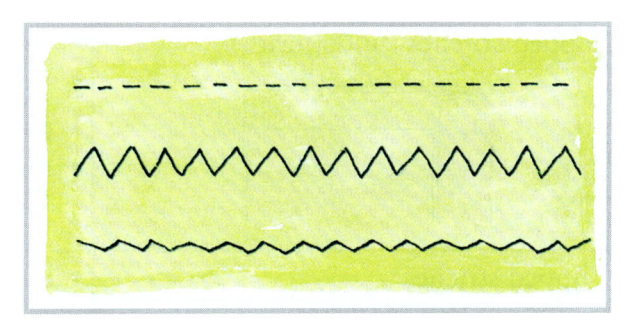

Straight stitch: used for seams and hems.
Zigzag: used mainly for seam finishing; a slight zigzag can be used for seams and hems on stretchy fabric.
Buttonhole stitch: machine buttonholes.
Machine embroidery

Choosing a sewing machine	Care of the sewing machine
• cost: compare models; compare the same models in different shops • buy a reliable brand from a reliable shop • buy an elaborate machine only if you intend to use its special features • check guarantee and after-sales service • ask for a demonstration	• carefully read care instructions which come with the machine • cover machine when not in use • clean dust from moving parts with small brush provided with the machine • oil occasionally (see care instructions for where to oil) • have the machine serviced and repaired by a sewing machine mechanic occasionally

Machine faults: possible causes

Fault	Possible causes
needle breaking	• needle has been put in back to front • needle is blunt and cannot get through fabric properly • fabric is too thick; sewing over zips or pins • presser foot loose, needle is hitting it
thread breaking	• machine not threaded properly • thread too fine or poor quality • faulty needle or needle in back to front
looped stitches (underneath)	• top tension too loose • top and bobbin threads not the same • presser foot not down • machine not threaded correctly
slipped stitches	• blunt needle • needle in back to front • needle is the wrong size for the fabric
puckered seam	• tension (top of bobbin) too tight • blunt needle • top and bottom threads are different • stitch too long (fine fabrics e.g. satin)

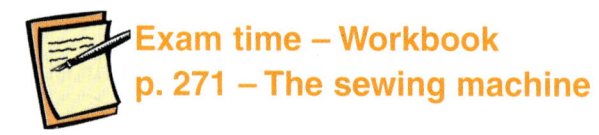

Exam time – Workbook p. 271 – The sewing machine

Seams

Seams are used to join pieces of fabric together to make clothing and other textile items. The flat seam pictured here is the most common type.

Flat seam

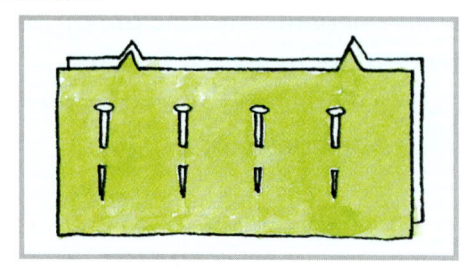

1. Pin right sides together, match notches

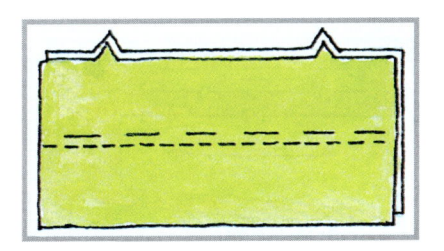

2. Tack, sew 1.5 cm from edge

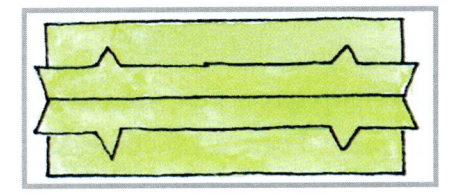

3. Press flat

Seam finishes

Raw edges may be finished in one of the three ways pictured here. This prevents fraying.

a) zigzag

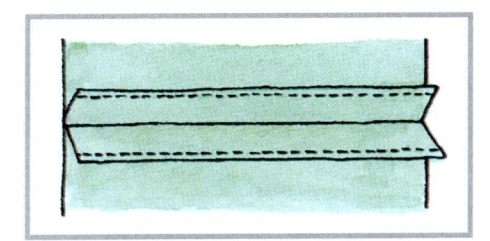

b) edge machining (fine fabrics)

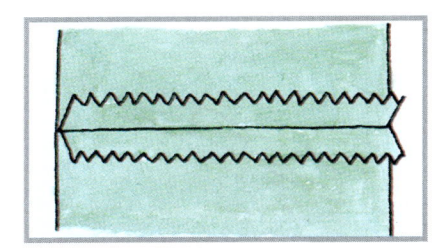

c) pinking shears (closely woven fabrics)

Choosing a fabric for home sewing

1. Choose a fabric that is easy to work with. Avoid slippery or stretchy fabrics as these are difficult to sew. Medium weight cotton is good for beginners.
2. Fabrics that have a nap (e.g. velvet) or a one-way design are more difficult to cut out properly and need more fabric.
3. Fabrics are sold in different widths. The wider the fabric you buy the less of it you will need.
4. Calculate the amount of fabric you will need using the grid on the back of your pattern envelope.
5. Check wash-care instructions on the fabric roll.
6. Buy anything else you need: a list of requirements or notions – matching thread, zip, buttons etc. – will be given on your pattern envelope.

Note: nap fabrics are smooth when brushed one way and rough when brushed the other.

Beginners should avoid fabrics with a nap or one-way design

Activity 36.3 – Workbook p. 273

ESSENTIALS FOR LIVING

Selvedge

Selvedge threads, also called warp threads, are strong threads which run up and down the length of the fabric. When cutting out pattern pieces, it is important that the selvedge runs down the length of the piece. This allows the finished garment to hang well.

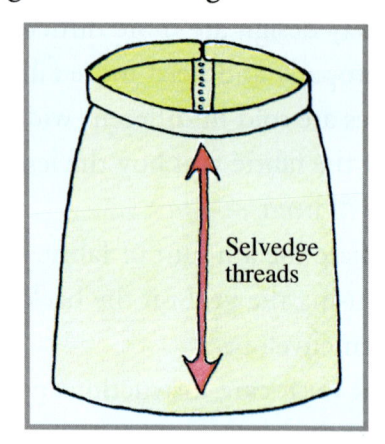

Bias: if fabric is cut on the bias (see diagram) it is stretchy. Strips of bias binding can be used to neaten arm holes, necklines or for piping.

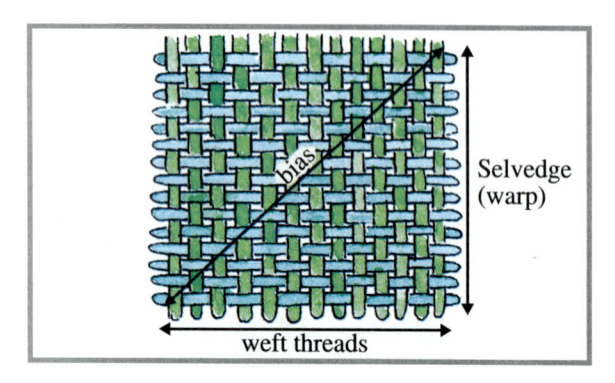

Cutting out

1. Arrange pattern pieces on the fabric. Straight-of-grain arrows ⟷ should run parallel ⇌ with selvedge edges. Place fold arrows ⊓ on fold. (Pattern instructions will suggest a layout.) 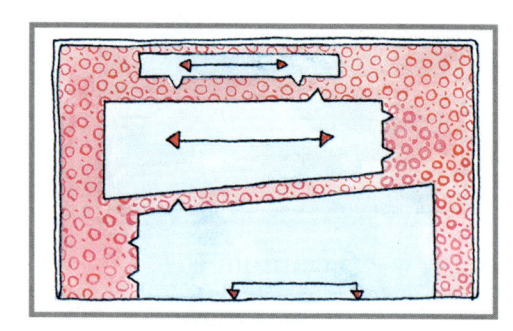
2. Pin in position.
3. Using a sharp scissors cut around each pattern piece. Leave a 1.5 cm seam allowance if not already allowed. (Most patterns include this allowance, in which case cut fabric to exact size of pattern pieces.)
4. Keep fabric flat on the table with one hand. Do not lift the fabric up to cut it. Cut notches outwards.
5. Transfer important pattern markings such as buttonhole location using (a) tailor tacks, and possibly (b) tailor's chalk and (c) a tracing wheel as pictured here.

a) tailor tacking

b) tailor's chalk

c) tracing wheel
Tailor tacking is always used; tailor's chalk and tracing wheel may also be used

288

Activity 36.4 – Workbook p. 274

Exam time – Workbook p. 276 – Seams

Now test yourself at *www.my-etest.com*

chapter 37

Fabric care

Caring for clothes

Before storing:
- mend
- remove stains, wash or dry clean
- fold knitwear, store flat in a drawer
- close zips and buttons; hang on shaped or padded hangers

CARE LABELLING

Care labels found sewn into clothing and household textiles carry five basic symbols:

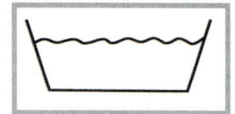

Symbol	Meaning
washtub	washing instructions
square	drying instructions
iron	ironing instructions
triangle	bleaching instructions
circle	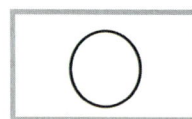 dry-cleaning instructions

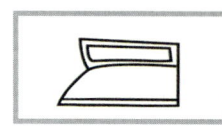

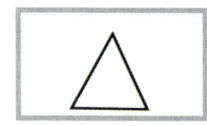

Washing instructions

There are three basic factors to consider when machine washing:

i) water temperature
ii) wash action: how fast the machine moves the clothes around during the cycle
iii) spin length: spin (full) or short spin

- water temperature is written inside the washtub symbol

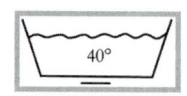

[washtub symbol with 40°]

- the bar symbol under the washtub tells you the correct wash action and spin length for the item you are washing

Bar symbols

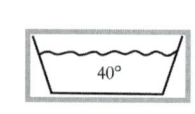

No bar: maximum/normal washing action and normal spin

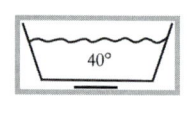

Single bar: medium/reduced washing action and short spin

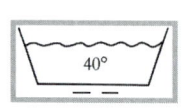

Broken bar: minimum washing action and wool cycle spin

FABRIC CARE

Tables like this one are found on washing detergent packets. They explain how fabrics can be washed and spun.

Textile/machine code	Machine wash	Hand wash	Fabric
95°	maximum wash in cotton cycle	hand-hot or boil, spin or wring	white cotton, linen; no special finishes
60°	maximum wash in cotton cycle	hand-hot or boil, spin or wring	cotton, linen, viscose; no special finishes; colours fast at 60°C
50°	medium wash in synthetic cycle	hand-hot, cold rinse, short spin	polyester-cotton mixes, nylon, polyester, cotton, viscose articles with special finishes, cotton-acrylic mixes
40°	maximum wash in cotton cycle	warm, spin or wring	cotton, linen, viscose where colours are not fast at 60°C
40°	medium wash in synthetic cycle	warm, do not rub; spin, do not wring	acrylics, acetate and triacetate including mixes with wool; polyester-wool blends
40°	minimum wash in wool cycle	warm, do not rub; spin, do not wring	wool, wool mixes, silk
(hand symbol)	hand-wash only	see individual care label	some pleated fabrics
(crossed out)	do not wash	do not wash	see individual care label

ESSENTIALS FOR LIVING

Drying instructions

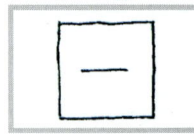

□ (—)	dry flat (wool)
□ (⌒)	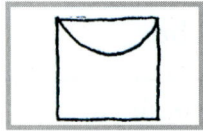 line dry
□ (|||)	 drip dry
□ (○)	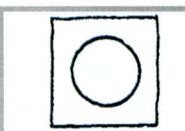 tumble dry
□ (⊗)	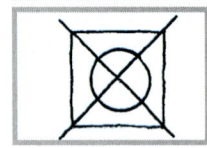 do not tumble dry

Ironing instructions

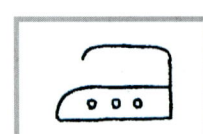

hot iron symbol	hot iron (cotton, linen)
warm iron symbol	 warm iron (wool, polyester, silk)
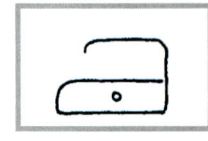 cool iron symbol	cool iron (nylon, viscose, acrylic)
crossed iron	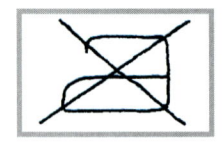 do not iron

Dry cleaning instructions

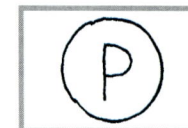

(A)   (P)

The letters tell the dry cleaner which chemicals to use.

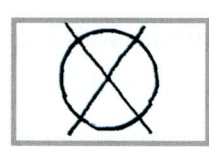

Do not dry clean

Bleaching

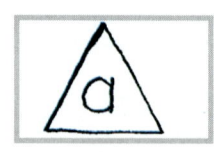

△ (a)	Bleach can be used
crossed triangle	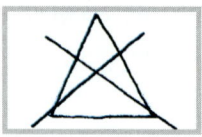 Do not bleach

Preparing a wash

- empty pockets
- repair any clothes which need mending
- remove stubborn stains before washing
- close zips and buttons
- sort clothes according to their care labels

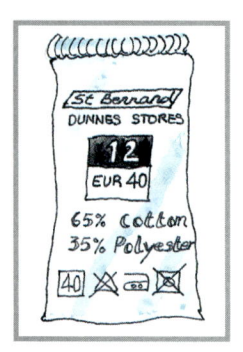

Care label

292

FABRIC CARE 37

Delicate fabrics
- wash by hand
- use a mild detergent such as Fairy Snow
- rinse twice
- squeeze to remove excess water
- roll in a towel to remove water
- drip or dry flat

Activity 37.1 – Workbook p. 277

Exam time – Workbook p. 278 – Care labels

Stain removal
- act quickly or stain will set
- do not rub stain into the fabric: blot or scrape off as much as you can
- use mildest treatment first e.g. soaking in cold water
- if you have to use a chemical stain remover, first test it somewhere on the garment that won't be seen
- remove stains before washing

Commercial stain removers (follow instructions given, use in a well-ventilated area)

Commercial stain removers for specific stains are available, such as Stain Devils: find out how much they cost.

293

ESSENTIALS FOR LIVING

Stain removal chart

Stain	How to remove it
protein stains: blood, egg, gravy	if stain is fresh soak in cold water, then wash; if stain is older soak in warm water with an enzyme (biological) detergent, then wash
chewing gum	freeze, pick off; you may have to use a grease solvent, e.g. benzene, or a Stain Devil
ink, grass	dab with methylated spirits and wash
chocolate	dab with glycerine and wash in hot water
tea, coffee, perspiration (sweat)	soak in warm water with an enzyme (biological) detergent, then wash
grease and oil	wash in hot water, dab with a grease solvent, e.g. benzene, if necessary
mildew (grey spots on cotton and linen)	whites: soak in a mild bleach solution colours: treat with hydrogen peroxide

Activity 37.2 – Workbook p. 279

Exam time – Workbook p. 280 – Stain removal

Detergents

Detergents such as Persil and Daz help remove dirt and stains from clothes. An emulsifier is an important ingredient in a detergent. Emulsifiers attach themselves to the dirt at one end and to the water at the other. This takes dirt off clothes and into the washing water.

Types of detergent

- powder or liquid detergents
- hand-washing (for delicates, e.g. Whip Express)

FABRIC CARE

- low-foaming (automatic powders for modern washing machines)
- biological (contain enzymes which break protein stains down at under 40°C; powder washes at other temperatures also)

Fabric conditioners

These are added during the final rinse. They (a) soften clothes, (b) reduce static and cling and (c) reduce wrinkling.

The washing machine

Guidelines for use

- wait, if you can, for a full load before using the machine

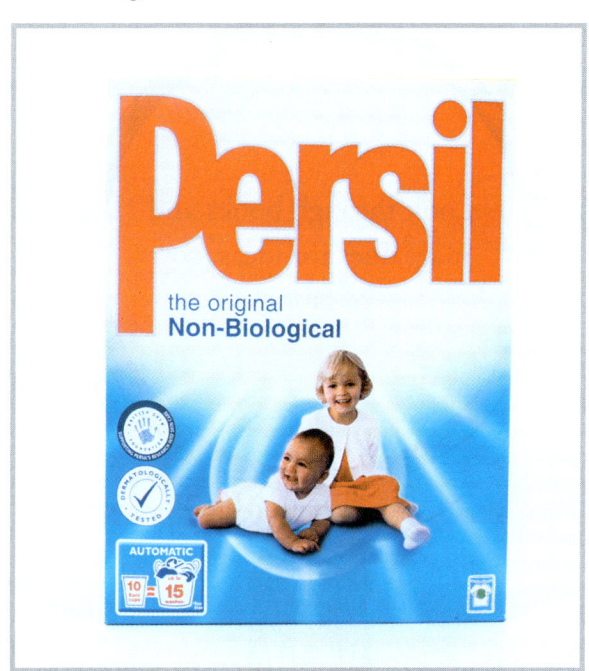

Low-foaming automatic washing detergent must be used in automatic machines

- use a low-foaming (automatic) detergent
- use the economy button if clothes are not very dirty

- know which programmes are suited to which fabrics

Drying

Clothes may be (a) line dried, (b) dried indoors on a clothes horse or (c) tumble dried.

Be careful when tumble drying, some clothes will shrink (wool) or discolour (synthetic underwear), if tumble dried. Tumble drying, while quick and great for removing creases, is expensive.

Ironing

Most modern irons are steam irons. They contain a heating element, water tank and thermostat which controls how hot the iron gets.

Guidelines for use:

1. Store iron sitting upright
2. Unplug to fill water tank and empty after use
3. Use pre-boiled water in hard water areas to prevent lime scale build up
4. Set iron to correct temperature (look at care label)
5. Iron on the wrong side unless you want a shine, e.g. linen tablecloths

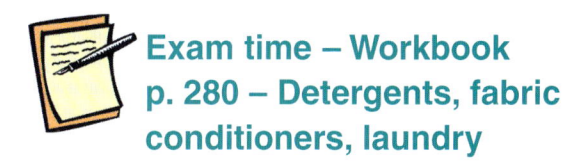

Exam time – Workbook p. 280 – Detergents, fabric conditioners, laundry

Now test yourself at *www.my-etest.com*

chapter 38

Practical work

As part of the Junior Certificate Textile Studies section, you must make two simple textile items:

- a household item
- an item of clothing

> Your finished pieces will not be examined directly but you may be asked questions about one or other of them in Section B, question 6 of your examination paper.

HOUSEHOLD ITEM

Sample brief

Design and make a household item. The item must be made from textiles and cost no more than €10 to make. You must be able to complete the item in approximately four weeks.

Analysis

What are the points that you must consider?
1. A textile must be used
2. The item must be for the home
3. It must cost no more than €10
4. It must not be too difficult for you to make; consider:
   - your sewing skills
   - the short time given to make the item

Research

Look through books and magazines for ideas, cost of fabrics

List possible solutions

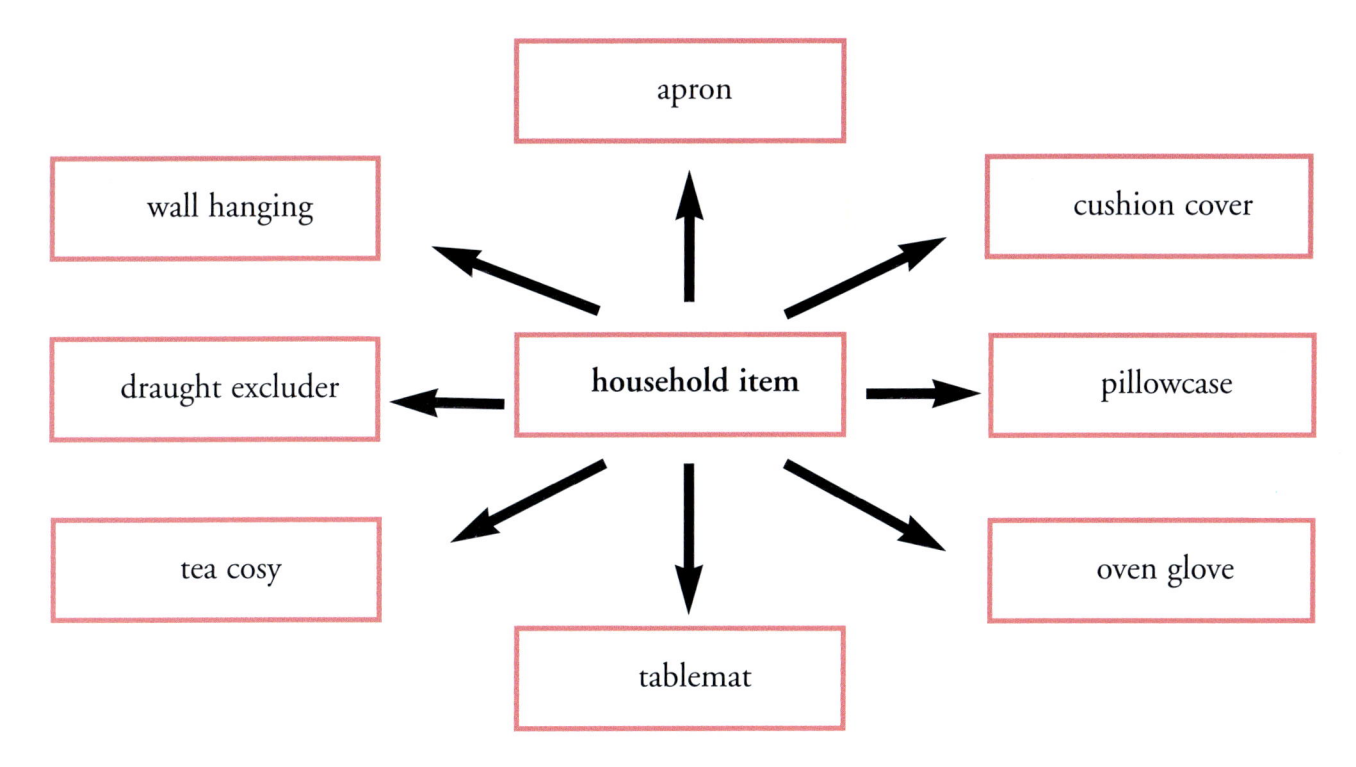

- apron
- cushion cover
- pillowcase
- oven glove
- tablemat
- tea cosy
- draught excluder
- wall hanging

(all connected to: **household item**)

Decide on a solution
Give a reason.

Action
Make the item.

Evaluate
Did you do everything you were asked to do in the brief?

Activity 38.1 – Workbook p. 282

Exam time – Workbook p. 284 – Making a household item

ITEM OF CLOTHING

The second textile item that you must make as part of the basic Junior Certificate Home Economics course is an item of clothing.

Activity 38.2 – Workbook p. 285

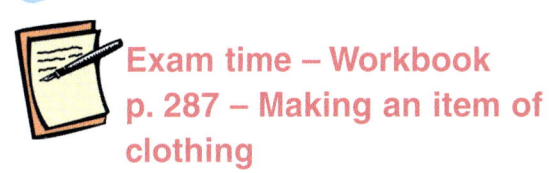

Exam time – Workbook p. 287 – Making an item of clothing

Unit 6

Options

chapter 39

Childcare option

Becoming a parent is a wonderful experience. However, even before conception, parenthood brings with it serious responsibilities.

A healthy pregnancy is not always guaranteed, no matter how careful the mother. A pregnant woman should:

DEVELOPMENT OF THE FOETUS IN THE UTERUS (WOMB)

(Pregnancy and birth: see page 195). While pregnancy generally lasts forty weeks, many babies are born prematurely (less than

- Avoid smoking, alcohol and drugs: these *will* harm the baby.
- Always check with your doctor before taking *any* medications.

- Avoid spicy foods such as hot curry, and strong tea or coffee.
- Danger of *listeria!* Avoid raw or lightly cooked eggs, unpasteurised cheese and cook-chill foods. Food-poisoning bacteria may be present – listeria is dangerous for the baby.

- Eat a well balanced diet: include fibre, protein, calcium and iron.

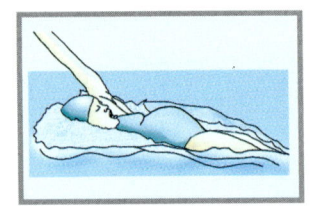

- Get plenty of exercise, fresh air, rest and sleep.
- Avoid too much weight-gain; 1.5–2 stone is ideal.

forty weeks) or are overdue (more than forty weeks). Rapid development takes place in the uterus during pregnancy.

Week 6

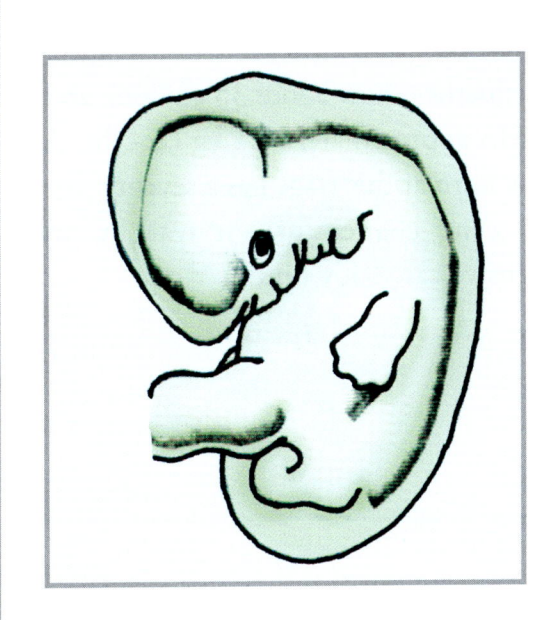

The embryo is now approximately 2.5 cm long. The heart is beating and arms and legs are beginning to form. It is very important that the mother does not allow harmful substances such as alcohol to pass to the baby at this stage, as all the organs are forming.

Week 12

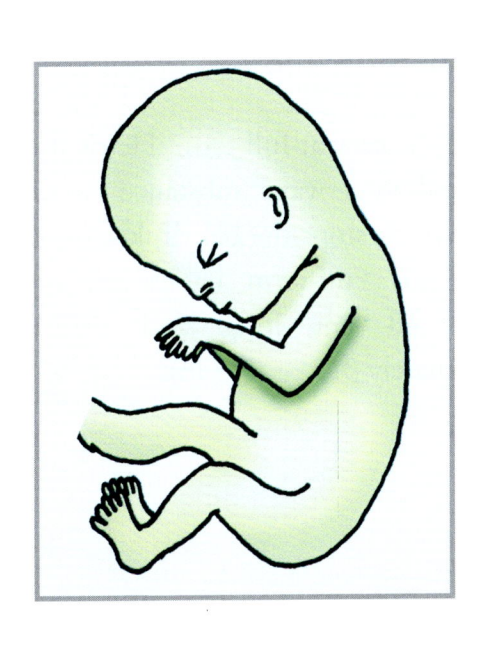

The foetus is now approximately 8.5 cm long. All the organs, both external (outside) and internal (inside), are formed. They must now mature.

ESSENTIALS FOR LIVING

Week 28

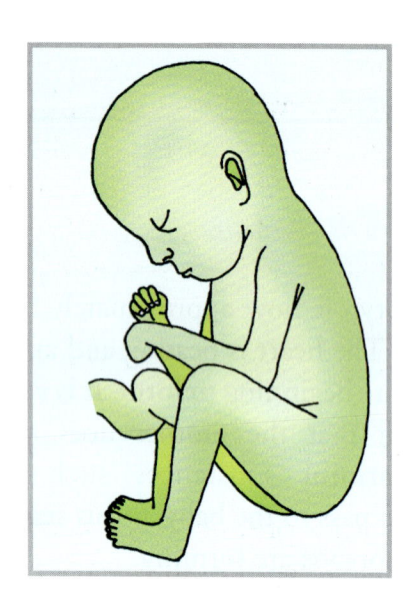

The foetus is now about 35 cm long and weighs approximately 1.1 kg (2.4 lb). A baby born at this stage has a seventy-five per cent chance of survival in an intensive care baby unit.

Week 38–40

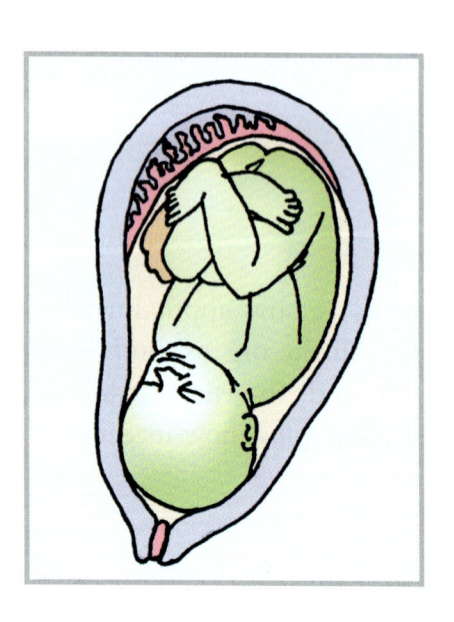

The baby is now at full term. He or she is covered in a waxy substance called vernix, which stops the baby's skin from drying out. Full-term babies are on average 48 cm long and weigh approximately 3.4 kg (7.5 lb).

Child development

Fresh air and exercise are vital to healthy physical development

Physical development

What it means:
- increase in height and weight
- developing gross motor skills: this means co-ordinating large body movements, e.g. crawling, walking, running, jumping etc.
- developing fine motor skills: this means co-ordinating small body movements, especially with the hands, e.g. feeding with a spoon, picking up small objects, holding a pen or pencil

Physical development is influenced by:
- heredity (what is passed on from parents, e.g. height, sporting ability)
- nutrition: a well balanced diet is vital for healthy physical development
- a healthy environment, e.g. plenty of sleep and fresh air
- opportunities the child gets, e.g. Tiger Woods began playing golf at three years of age!

Children should be given opportunities to learn and develop to their full potential

Intellectual and language development

What it means:
Intellectual and language development are very closely linked.

Intellectual development:
- reasoning (e.g. if I eat my dinner, I will be allowed out to play)
- problem solving (e.g. if I make the base of this block tower wider it won't keep falling over)
- concept formation (a pig is a pink animal with a curly tail)
- attention and memory

Language development:
- listening and understanding
- talking
- reading
- writing

Intellectual and language development is influenced by:
- heredity
- environment

A child who is encouraged and stimulated, talked to and listened to, read stories, exposed to interesting activities such as cooking, nature walks etc. has a much better chance of reaching their potential than a child who experiences few of these things.

ESSENTIALS FOR LIVING

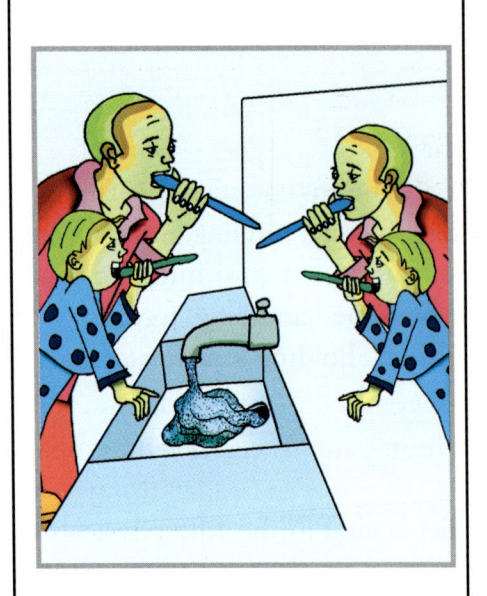

Children learn much through imitation

Social and emotional development

Social development

What it means:

Social development is how a child learns the norms (acceptable behaviours) of their society and how to relate well to others. Norms of behaviour: using the toilet, wearing clothes etc.

Relating well to others: sharing, turn-taking etc.

Emotional development

What it means:

Emotional development means learning to cope with feelings and development of self-esteem and self-image (how you see yourself).

Social and emotional development is influenced by:

- heredity
- environment

A child who grows up in a secure, loving environment, who is encouraged and praised, disciplined consistently and fairly, is more likely to have high self-esteem and a good self-image than a child who experiences few of these things. Children develop socially by imitating those around them and mixing with other children.

CHILDCARE OPTION 39

Stages of child development from birth to three years

Stage	Physical development	Intellectual and language development	Social and emotional development	Suitable toys and activities for age group
Birth	• at birth reflexes are present e.g. grasping reflex (see below) • no head control • can see 25 cm from nose	• startled by loud noises	• cries to make needs known	• plenty of physical contact, breast-feeding helps mother-baby bond • constantly talk in a soothing voice to the baby • play music to the baby
6 weeks	• when lying on front can lift head up for a few seconds	• recognises familiar voices • gurgles and coos	• smiles at about 6 weeks • looks at parent's face intently when feeding • cries to make needs known	• things to look at e.g. mobiles • plenty of close physical contact • talk to and interact with the baby constantly • music
3 months	• body is now uncurled, legs and arms are outstretched • kicks vigorously • head steady • plays with own hands	• chuckles • turns head towards sounds	• gets excited at voices, sounds e.g. bath water running, sight of bottle	• things to kick e.g. baby gyms (home made are just as good) • music

305

Stage	Physical development	Intellectual and language development	Social and emotional development	Suitable toys and activities for age group
6 months	• sits with support turning head from side to side to look around • holds hands up to be lifted • when standing can take weight on legs, enjoys bouncing up and down • puts everything into mouth	• squeals aloud in play • babbles tunefully e.g. *a dah aroo goo ga*	• still friendly towards strangers (fear of strangers develops around 7 months)	• picture books • bath toys • baby walker (gives independence) • toys to bang e.g. drum • activity centres
9 months	• can sit alone • rolls along on floor • attempts to crawl, sometimes succeeds • can stand holding on • passes toys from hand to hand • uses finger to point	• shouts to attract attention • understands simple words e.g. *bye bye* or *no* • babbles *da da*, *mam ma* • playful vocal sounds e.g. 'Jack had a bad cough' – will pretend to cough • may find object hidden under a cup or cushion, for younger children out of sight is out of mind	• can hold and bite a biscuit • clings to known adult, fears strangers • plays peek-a-boo • claps hands	• songs and action rhymes • encourage crawling, give toys that roll • give finger foods

CHILDCARE OPTION **39**

Stage	Physical development	Intellectual and language development	Social and emotional development	Suitable toys and activities for age group
12 months	• some babies 'bear walk' before walking • may walk alone • can crawl upstairs (care is needed as cannot come down safely) • can pick up small objects e.g. crumbs • points to what he/she wants	• knows own name • understands simple instructions e.g. 'give the ball to daddy' • understands the use of everyday objects e.g. hair brush • waves bye bye	• helps with dressing e.g. holds foot out for shoe • likes to be constantly in sight of familiar adult • can bring spoon to mouth, may tip it upside down and spill food	• picture books • push-along toys • fetching games e.g. get teddy • songs and rhymes
18 months	• walks well carrying a toy • runs carefully with eyes fixed on the ground; will therefore bump into things • climbs onto adult chair and then turns around • makes scribbles and dots on paper • turns several pages of a book around at once • starts to show left or right handedness	• recognises familiar faces in photos • uses 20+ recognisable words • imitates everyday activities e.g. feeding a doll, brushing the floor	• can feed him/herself • can take off shoes, socks and hat • will play alone but still likes adult to be near at hand • tells adult when nappy is soiled and wants it removed	• dancing to music • pull/push-along toys • crayons and paper • bricks • shape sorters • encourage child to 'help' with household chores

Stage	Physical development	Intellectual and language development	Social and emotional development	Suitable toys and activities for age group
2 years	• can run and avoid obstacles • walks upstairs and (often) down, holding rail or wall, two feet to a step • throws a ball without falling over • sits on a tricycle, but cannot use the pedals • can remove paper wrapping from sweet • holds pencil and scribbles • turns pages singly	• 50+ recognisable words • 2- or 3-word sentences • refers to self by name • constantly asking the names of objects and people • can join in nursery rhymes and songs • enjoys pretend play	• spoon-feeds without spilling • no comprehension of danger e.g. will run out onto road • has temper tantrums (usually 1–2 per day) • will not yet share willingly • plays near other children but not with them • jealous of other siblings getting attention from important adults	• jigsaws • simple story books • play dough • painting • threading toys • musical instruments • tricycle • props for pretend play e.g. a cash register for shop

CHILDCARE OPTION 39

Stage	Physical development	Intellectual and language development	Social and emotional development	Suitable toys and activities for age group
3 years	• rides a tricycle using the pedals • kicks ball forcibly • can avoid obstacles when running • can cut out with safety scissors	• can write X, V, H and T • draws a person with head and 1–2 other parts / features • may know names of colours • covers entire page with paint often uses only one colour • asks many questions: 'what', 'where', 'who' • loves stories, demands favourites over and over • counts to 10 or more but has little understanding of number	• eats with a fork and spoon • can pull pants up and down • toilet trained • plays with other children • understands sharing • shows affection for other siblings	• story books • educational computer games • crayons, paint and paper • construction toys e.g. Lego • sand and water play • cooking activities • encourage children to 'help' with household chores • large outdoor equipment e.g. slide • make-believe toys

FEEDING

Babies feed on milk alone for the first four months of life. Babies may be bottle- or breast-fed.

Making up a formula feed

1. Wash hands carefully.
2. Scrub all baby-feeding equipment thoroughly with a bottle brush, washing up liquid and hot water; rinse.
3. Sterilise using one of the following four methods:

ESSENTIALS FOR LIVING

a) Boil everything in a saucepan with the lid on for three minutes.

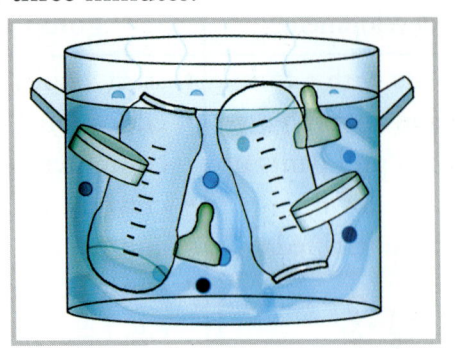

b) Use a steam sterilising unit which clicks off automatically when the bottles are sterilised; follow manufacturer's instructions exactly.

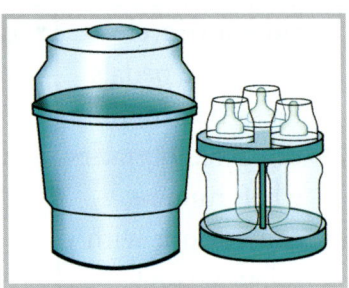

Advantages and disadvantages of breast- and bottle-feeding

Breast-feeding	
Advantages	**Disadvantages**
• nutrients are in the correct proportions for the baby's needs • antibodies are passed from the mother to the baby; breast-fed babies are better able to fight disease • breast milk is totally sterile; no risk of food poisoning • breast-feeding encourages bonding • breast milk is free and always available • breast-feeding helps mother regain her figure • breast-fed babies are less likely to be overweight, suffer from allergies, have colic or constipation	• strain on mother who must do all the feeds • other members of the family, e.g. father, may feel left out • during the first week breast-feeding can be very sore • there may be a lack of breast-feeding facilities while out and about
Bottle-feeding	
Advantages	**Disadvantages**
• formula milks are now available which copy breast milk exactly • bottle-feeding gives the mother more freedom; night feeds can be shared between parents	• badly made-up feeds can be dangerous; never put extra scoops of formula into the bottle, it can cause the baby to dehydrate or become very constipated • bottles which have not been sterilised properly can give the baby food poisoning and make him or her very ill • formula milks are expensive and making up feeds is time consuming

c) Use a cold water sterilising tank. Sterilising fluid or tablets are added to cold water; bottles etc. must be completely covered with fluid for thirty minutes.

d) Sterilisers for use in the microwave are also available; follow instructions exactly.
4. *First*, pour the exact amount of cool *boiled* water into the sterilised bottles.
5. *Second*, add the correct number of *level* scoops of dried formula milk (the pack will tell you how many scoops to add).
6. Replace the teat and lid and shake well.
7. If you make up more than one feed, store the others in the fridge; reheat carefully in a jug of boiling water (microwave reheating can be dangerous as the outside of the bottle may be cool and the inside hot: shake well).
8. Hold the baby close when feeding. A baby should never be left alone to feed with a propped-up bottle.
9. Empty leftover feed out; rinse and wash bottle ready for sterilising.

Weaning

Babies are usually ready for weaning at around sixteen weeks. Babies should not be weaned too early as their digestive systems are not developed enough to cope with solids. Breast- or bottle-fed babies have an instinct to feed, but eating solid food takes practice and babies may appear to be spitting out food even when they are not.

Baby rice and puréed fruits and vegetables are very suitable first foods. Never add salt or sugar to feeds and avoid foods containing gluten (wheat) as it can be difficult to digest.

Commercial baby foods

Commercial baby foods have improved greatly in recent times and now there is a huge variety on the market. Such foods are very useful in emergencies but are expensive if used all the time.

CARE OF THE TEETH

Never give babies juice in a bottle. It leads to bottle rot

A baby's teeth start to appear around six to seven months. Even though baby- or milk-teeth only last until approximately age seven it is very important that they are well cared for.

- never give babies juice to drink from a bottle: this leads to bottle rot; even young babies can drink small amounts at a time from a cup or spoon
- do not allow babies and toddlers to fall asleep with bottles of milk in their mouth: this leads to bottle rot
- once baby-teeth appear, clean them using a cotton bud or later a soft toothbrush; teach toddlers to brush their own teeth as soon as they are able
- do not bribe children with sweets or dip soothers in sugar: you are encouraging a sweet tooth and may cause tooth decay
- provide children with plenty of calcium-rich foods such as cheese, yoghurt and milk

CHILDREN WITH SPECIAL NEEDS

Children with extraordinary abilities also have special needs

All children have needs, for example the need for food, shelter, clothing, love, security and understanding. Children with disabilities or extraordinary abilities have these same needs together with some additional ones.

> A *disability* is something that can restrict the individual or make some things more difficult to do.

Children with special needs may have:
- moderate to severe learning difficulties, e.g. Down's syndrome
- sensory impairment, e.g. hearing or visual impairment
- physical impairment, e.g. cerebral palsy, diabetes or coeliac
- speech and language difficulties
- emotional and behavioural difficulties, e.g. attention deficit hyperactivity disorder (ADHD)

- specific learning difficulties, e.g. dyslexia
- extraordinary ability/abilities

Always treat a child with special needs as the individual they are; ensure that the child is considered first and the special need second.

Some common disabilities explained

It must be pointed out that only a very basic explanation is offered here. All of the disabilities below are very complex.

Name	Cause	Effects	Special needs
Asthma	The airways of the lungs become narrowed. Allergies, infection, exercise, the weather or emotional upset may bring on an attack.	Breathing becomes difficult. If an attack is severe, child may become anxious and afraid. Asthma attacks can be fatal.	Child must avoid things or situations they know bring on attacks. An inhaler is used to widen the airways again and bring breathing back to normal.
Cerebral Palsy	The part of the brain that controls movement and posture is damaged or fails to develop. This may be caused by: • complications during pregnancy • accident or injury during or after birth	Movement of limbs is impaired or in other cases balance may be affected.	Physiotherapy, Speech therapy, Occupational therapy: helping person develop new life skills
Coeliac	See page 30		
Cystic Fibrosis	Hereditary disease (passed on in the genes)	Body mucus is very thick • breathing problems • severe chest infections • digestive problems	• antibiotic therapy for lung infections • lung massage • digestive enzymes taken at every meal

Diabetes	See page 31		
Dyslexia	Dyslexia is inherited. It may be inherited from an uncle, aunt or grandparent.	Dyslexia is a language-based learning disability in which a person has trouble reading. It is estimated that 15–20% of people have some form of learning disability.	Studies show that individuals with dyslexia often have above average intelligence. Specific language learning techniques are used to overcome the problem: the earlier the better. Unfortunately few teachers are sufficiently trained in this area.
Down's syndrome	Abnormality on the twenty-first chromosome.	• mental handicap • physical disability • frequent illness; prone to chest infections	A stimulating environment and a positive attitude will help the child reach his/her full potential.
Hearing impairment	Can be hereditary or caused by infections such as rubella during pregnancy or meningitis in the child after birth.	Degree of hearing loss can be from slight hearing loss to profound deafness.	Ear implants are becoming more common. Speech therapy, sign language.
Spina Bifida	True cause not known; both genetic and environmental factors thought to be involved.	Gap in the bones of the spine exposing the spinal cord. Several different types of Spina Bifida; ranges from little or no disability to paralysis.	Children should be helped to partake in all the activities usual for their age group.
Visual impairment	Hereditary or caused by infections during pregnancy or birth, e.g. rubella.	Individual may be blind or partially sighted.	Many visually impaired children, if given good resources and support, do very well in mainstream schools.

IRISH FAMILY LAW

The structure of the Irish family has changed dramatically over the past number of decades. There have been large increases in:
- single-parent families and children born outside marriage
- marital separations and divorces
- cohabiting couples who, for one reason or another, decide not to marry
- families where both parents work

Activity 39.1 – Workbook p. 293

As a result of all this change, Irish family law has had to be changed and updated. Some of the most common laws are listed and briefly explained below.

1. **Family Law (maintenance of spouses and children) Act 1976:** this Act requires that one spouse support the other and his or her children.
2. **Family Law (protection of spouses and children) Act 1981:** this Act gives the circuit or district courts the power to grant barring or protection orders.
3. **Domestic Violence Act 1996:** extended safety, barring and protection orders to non-married persons, meaning a boy- or girlfriend could be barred from the family home.
4. **Status of Children Act 1987:** did away with the word *illegitimacy* as a way of describing children born outside marriage. It also allows unmarried fathers to apply for guardianship of their children, provided they undertake a blood test to establish paternity.
5. **Family Law (divorce) Act 1996:** allows divorce and remarriage in this country.
6. **Children's Act 1997:** recognises natural fathers as guardians and allows children's views to be considered when guardianship, access and custody matters are being decided.

Activity 39.2 – Workbook p. 293

In addition to the law there are a number of organisations available to families experiencing difficulties.

- **Family mediation service:** this is a state-run service which helps couples who have decided to separate to sort out difficult issues such as custody of children, maintenance etc.
- **Irish Catholic Marriage Advisory Council (CMAC):** provides a counselling service to couples whose marriages are in difficulty.
- **Women's Aid:** provides temporary accommodation and counselling for women victims of domestic violence and their children.
- **Gingerbread:** a support group for single parents.

Activity 39.3 – Workbook p. 293

CHILDCARE PROJECT

You will not be asked questions on the Home Economics written paper about the childcare option, instead the childcare option, like each of the other two options, is assessed through project work. Your project is worth fifteen per cent of your final Junior Certificate mark at both ordinary and higher levels.

For this project you must select a *topic* and then research and prepare a written report on it. The written report should *not* be more than *1500 words* (approximately five full typed pages or nine written pages). You will not receive good marks for large amounts of writing copied closely from books and leaflets. Read the section below for ideas on how to research and present a good project.

Stage 1: Decide on a topic
Some possible topics:
- feeding babies and young children
- a child developmental study (where you choose a child and follow his or her development over a period of time)
- experiences of being a teenage mother or single parent
- a baby-sitter's handbook
- a guide for new parents
- make a child's toy as part of an overall study on child development
- a study of childhood illness and immunisations
- an investigation into childcare facilities in your area
- children with special needs
- safety and accident prevention in the home, first aid for babies

Stage 2: Make a contents page
Include:
- Aims
- Planning
- Research
- Presentation of main findings
- Evaluation and conclusions
- Bibliography and sources of information

Stage 3: Plan your project
See sample spider plan on page 318.

Stage 4: Research your topic
A good project will have two kinds of research:
1. *Primary research:* with this type of research you gather information yourself. You can do this by using questionnaires, interviewing people, observing children at play and recording what you see, visiting a play school and recording what you see etc.
2. *Secondary research:* with this type of research you read and use other people's primary research. Secondary research is taken from books, the Internet, magazines, information leaflets etc. It is very important that you do not copy straight from these sources into your project, as you will lose marks.

Stage 5: Present your findings
What are the most important things you

found out? You could present some of your findings as bar or pie charts. Some findings would not be easily presented this way and you may just want to list them.

Stage 6: Evaluation and conclusions

In this section you should review your aims and decide how well you achieved them. Explain why you feel you achieved or did not achieve them. If you were to do this project again list the changes you would make.

Stage 7: Bibliography and sources of information

In this section you should list where you got your information. Include books, magazines, leaflets etc.

One great source of information on health issues is the *Health Promotion Unit*. They have a huge range of leaflets and booklets on many of the issues you will be investigating as part of this project. Which health promotion office you write to or telephone will depend on which area of the country you are living in. Pick the nearest one from the list opposite.

Dublin Health Promotion Unit, Department of Health, Hawkins House, Dublin 2. (01) 670 7987

North-eastern Health Promotion Unit, Railway Street, Navan, Co. Meath. (046) 76401 / 76407

Midlands Health Promotion Unit, Third Floor, The Mall, William Street, Tullamore, Co. Offaly. (0506) 46730

Mid-western Health Promotion Unit, Park View house, Pery Street, Limerick. (061) 316655

North-western Health Promotion Unit, Main Street, Ballyshannon, Co. Donegal. (072) 52000

South-eastern Health Promotion Unit, Dean Street, Kilkenny. (056) 51702

Southern Health Promotion Unit, Eye, Ear and Throat Hospital, Western Road, Cork. (021) 492 3480

Western Health Education Centre, Shantalla, Galway. (091) 546005

ESSENTIALS FOR LIVING

Sample spider plan

Activity 39.4 – Workbook p. 294

Interview a mother on her experiences of breast-feeding, bottle-feeding and weaning her baby: prepare a set of interview questions before the interview, record the interview on tape, write out what was said during the interview and present this with your research

or

Ask a number of mothers to fill out a questionnaire you have designed on breast- or bottle-feeding and/or weaning

- Present findings
- Include bibliography and sources of information
- Present project nicely: check spellings, use photos and diagrams to brighten it up
- Look up breast-feeding, bottle-feeding and weaning on the Internet

To investigate infant feeding from birth to one year

- Have a contents page to include aim, planning, research, presentation of findings, evaluation and conclusions, bibliography
- Write to baby food companies e.g. SMA for information on infant feeding
- Present any Internet articles, letters, leaflets you received from companies as an appendix at the back of the project, not in the main body of the project
- Write to the health promotion unit for leaflets on breast-feeding, bottle-feeding and weaning: read these and summarise main points
- Prepare and cook some weaning dishes. Compare to similar bought varieties under the headings: nutritive value, taste, cost, convenience etc.

Now test yourself at *www.my-etest.com*

318

chapter 40

Design and craftwork option

Craft has a long, rich history in Ireland. In the past craft items, such as knitted jumpers, baskets and woven fabrics, would have been made in every household. Many such crafts were made firstly as functional items and only secondly as items of beauty. Once these items became mass produced in factories many Irish traditional crafts all but died out.

Nowadays, while many Irish craft items are functional, how the item looks is very important.

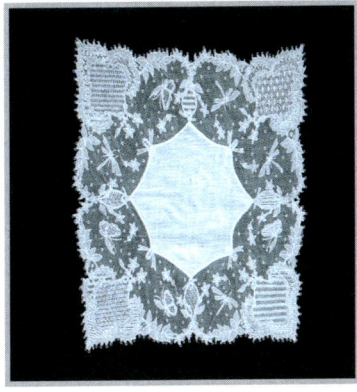

Carrickmacross lace

Waterford crystal

Beleek china

The Irish Craft Council

The Irish Craft Council, which was formed in 1971, is the national agency for the development of the craft industry in Ireland. The main aims of the council are to:
- develop markets for Irish crafts at home and abroad
- encourage excellence in the Irish crafts industry
- provide training courses in various Irish crafts
- provide information to those interested in the craft industry
- run exhibitions to promote Irish crafts

The Irish Crafts Council currently runs three full-time courses:
- jewellery making
- pottery
- blacksmithing and forging skills

ESSENTIALS FOR LIVING

The council is also involved with a number of short courses nationwide.

> For a broad view of the craft industry in Ireland today why not visit the Craft Council's website on *www.ccoi.ie* or write to them at Crafts Council of Ireland, Castle Yard, Kilkenny. Telephone (056) 61804

Activity 40.1 – Workbook p. 295

DESIGN AND CRAFTWORK PROJECT

You will not be asked questions on the Home Economics written paper about the design and craftwork option, instead the design and craftwork option, like each of the other two options, is assessed through project work. Your project is worth fifteen per cent of your final Junior Certificate mark at both ordinary and higher levels.

> **Brief**
> For this project you must design and make a simple, cost-effective craftwork item from a *textile*.

The design and craftwork folder

Together with the craft item you must present a written folder, which must contain:

1. A table of contents:
   - Analysis of brief (points 2, 3 and 4 below)
   - Solution (points 5, 6, 7 and 8 below)
   - Evaluation (points 9 and 10 below)
2. A statement of what you have been asked to do
3. A list of the most important things you have to consider
4. Evidence that you investigated a number of different crafts before deciding on one
5. Background information on the craft you have chosen; you could look on the Internet for this, or in books and magazines from your Home Economics room or local library
6. A plan of work, similar to this sample plan:
   - *Week 1:* Make a list of all the materials and equipment I need and go and buy them
   - *Week 2:* Practice my craft on old scraps of fabric
   - *Week 3:* Complete sample drawings and designs; choose one, etc.
7. A list of all materials and equipment you used for making the craft item, with costings
8. A description with diagrams, sample designs, patterns etc. of exactly how you made your craft item
9. An evaluation of the craft item: does it meet the brief? Why?
10. Proposed modifications: what would you change if you were to do it again, and why

The craft item itself:

Stage 1: research and choose a craft; for this you will have to consult with your teacher and carry out research in books, magazines, craft shops etc.

DESIGN AND CRAFTWORK OPTION **40**

Some possibilities are:
- patchwork
- rug making
- appliqué
- embroidery
- cross-stitch
- lace making
- crochet
- knitting
- quilting

Cross-stitch

Patchwork

Simple appliqué

Stage 2: decide what to make that would best suit your craft.

Some possibilities are:
- cushion covers (suits most of the crafts above)
- pictures (appliqué, embroidery, cross-stitch, lace making)
- rugs (rug making)
- wall hangings (patchwork, appliqué, embroidery)

Stage 3: practice your craft.

You should practice your craft on old scraps of material: keep these and present them in your design folder.

Stage 4: finalise your design on paper.

Stage 5: make your craftwork item.

Now test yourself at *www.my-etest.com*

chapter 41

Textile skills option

This option gives you the opportunity to make an item of clothing, and to learn about careers opportunities in the textile industry, for example as a clothing designer.

Using patterns and basic garment construction

Before beginning this chapter please revise Textile studies, Chapter 36, page 280. Pay particular attention to the following:
- using a pattern envelope
- choosing fabric for sewing
- pattern layouts
- transferring pattern markings
- cutting out pattern pieces

Taking body measurements

Before you buy a pattern or fabric you will need to know your correct body measurements.

General guidelines
- use a tape measure that has not been stretched and do not pull it too tight as it will not give the correct reading
- remove outer garments, such as jumpers, so that you are not measuring over too much bulk
- ask a friend to help: it is very difficult to take your own measurements correctly
- check all measurements twice: it is easy to make mistakes

For most garments, you need to measure your:

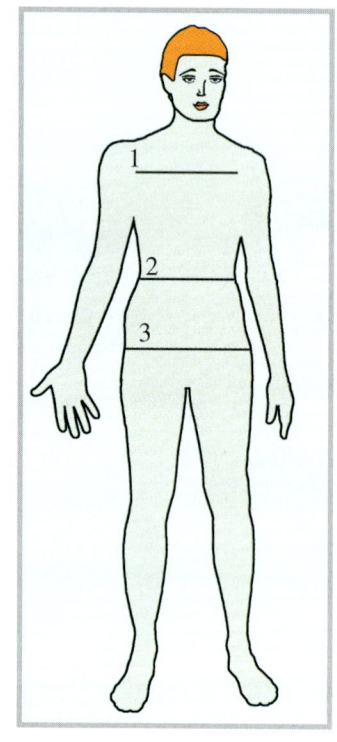

1. chest (male) or bust (female)
2. waist
3. hips

For all of these, measure around the fullest (widest) part. To find your natural waistline, tie a piece of string around your waist: where it rests comfortably is your natural waistline.

For some garments you may need additional measurements:

TEXTILE SKILLS OPTION 41

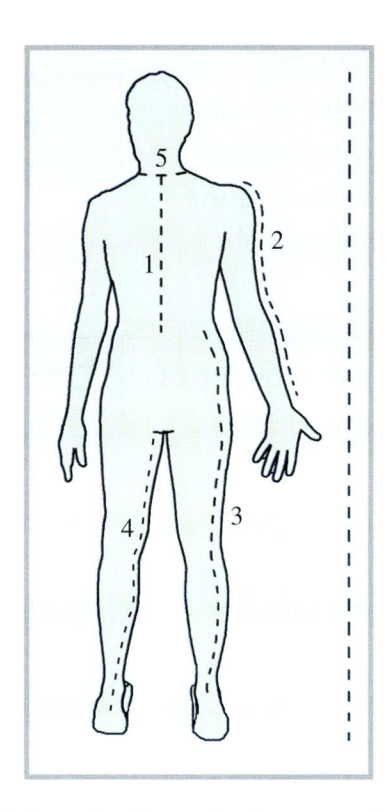

1. Back length: from the bone at the bottom of your neck to your natural waistline.
2. Sleeve length: from your shoulder to your wrist keeping your elbow bent.
3. Outside leg: from your waist down to where the bottom of the hem will be.
4. Inside leg: from your crotch to where the bottom of the hem will be.
5. Neck width: measure around the neck and add 1.2 cm to the result (so that collars won't be too tight).

Activity 41.1 – Workbook p. 296

MAKING CLOTHES

Choosing a pattern

Several different companies, such as Style, Simplicity and Burda, manufacture commercial patterns. These companies display their patterns in pattern catalogues, which are available for you to look at in textile shops. Most of these companies manufacture patterns that are suitable for beginners; these are often found in a special section of their catalogues.

Most patterns include more than one style (view) and a range of sizes.

Inside the pattern envelope

Inside the pattern envelope you will find two things:
- the actual pattern pieces
- a set of instructions

Pattern pieces

Inside the pattern envelope you will find the pattern pieces for all the views pictured on the pattern envelope. Pattern pieces will be printed out on a number of very large pieces of tissue paper. You will have to study the pattern instructions to find out which pieces are needed for the particular garment you are making.

ESSENTIALS FOR LIVING

Pattern pieces will have the following information printed on them:
- the number of the pattern piece
- the name of the pattern piece, e.g. *skirt back*
- the following very important markings:

Notches
These show where the garment pieces should be joined. Cut notches outwards.

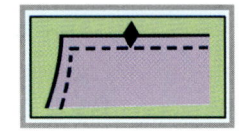

Straight grain
This arrow should be running along the selvedge (warp) of the fabric.

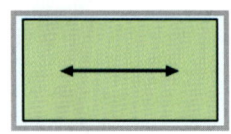

Fold line
This line should be placed along the fold of the fabric.

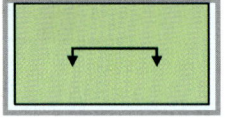

Cutting line
Most patterns have a cutting line which is 1.5 cm outside the stitching line; this is where you cut out. Note: some patterns do not have a cutting line: you must allow the 1.5 cm seam allowance yourself.

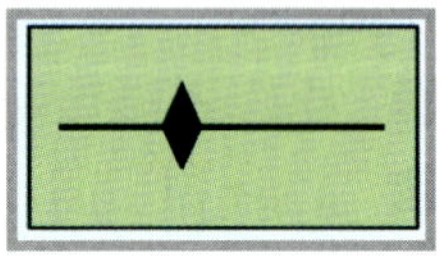

Balance marks
These show important points on the garment, such as the centre of the garment, the armhole.

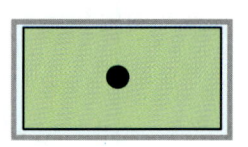

Construction marks
These show the position of darts, pockets, pleats etc.

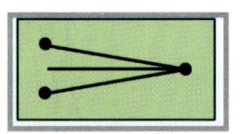

Stitching line
Machine along this line.

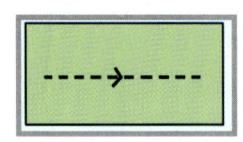

Alteration lines
It is at these lines that pattern pieces can be made longer or shorter (see below).

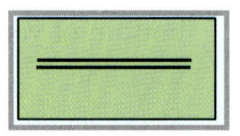

Button positioning marks
These should be marked with a tailor tack to show where buttons are to be placed.

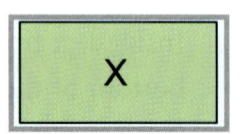

Buttonhole positioning marks
These show where buttonholes should be sewn. Notches, balance marks, button and buttonhole positioning marks should all be transferred from the tissue pattern to your fabric using tailor tacks (see Chapter 36).

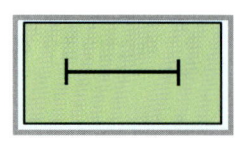

Stitching lines and construction marks may be transferred to fabric using a tracing wheel and paper.

Activity 41.2 – Workbook p. 296

Instruction sheets

Usually contain the following information:
- a sketch of the back of the garment(s)
- a labelled sketch of all the pattern pieces included
- a list of the pieces needed for each view
- instructions for altering the pattern, e.g. making it longer
- a sketch showing how pattern pieces can best be laid out on fabrics of different widths
- cutting instructions
- step-by-step instructions on how to make the garment

Activity 41.3 – Workbook p. 296

Altering patterns

While small adjustments can be made on the garment itself, for example letting out a seam slightly, larger adjustments must be made on the pattern pieces before placing them on the fabric.

To lengthen pattern pieces

Example: you wish to add 3 cm to the length of a pair of trousers.
1. Cut the pattern piece between the alteration lines.
2. On another piece of paper draw two parallel lines 3 cm apart.
3. Lay one pattern piece on each of the parallel lines. Stick in place using masking tape.
4. Redraw the cutting line. Trim excess paper.

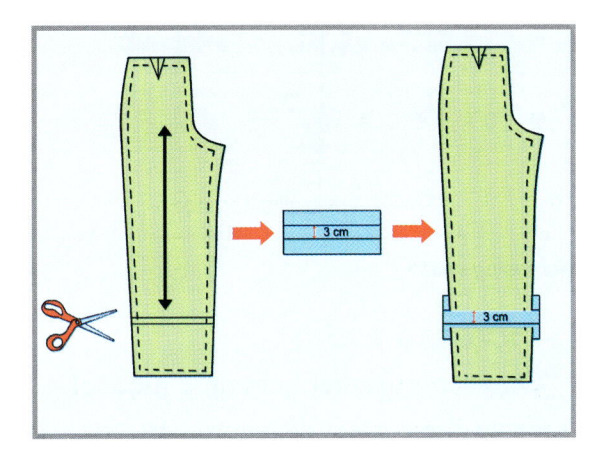

To reduce the length or width of a pattern

1. Fold the pattern piece on the alteration lines. The fold should measure half the width that you need to reduce. Tape into place and redraw the cutting line.

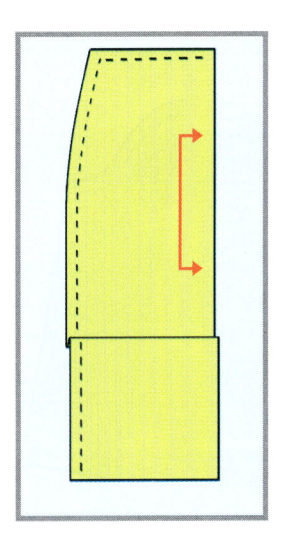

ESSENTIALS FOR LIVING

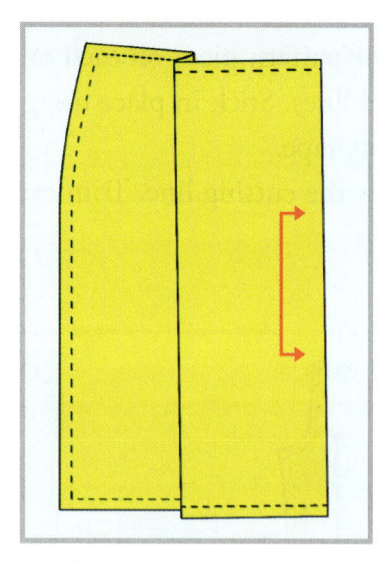

Decreasing width

To increase width

1. Draw two parallel lines on a piece of paper the correct distance apart; for example if you want to increase width by 3 cm the lines should be 3 cm apart.
2. Tape the cut pattern piece along the two parallel lines. Redraw the fitting lines etc.

Remember if you increase the width of one part of the garment the pieces it is attached to must also be increased to allow pieces to join together.

Laying out pattern pieces

(See also Chapter 36, textbook and workbook.)

Your pattern instructions will include diagrams showing how your pattern pieces should best be laid out. Layouts will be different for different fabric widths.

Generally

- press fabric on the wrong side before beginning pattern layout
- fold the fabric right sides and selvedge edges together
- pattern pieces are generally placed on the fold or laid on the fabric using the straight-of-grain lines (see next page)

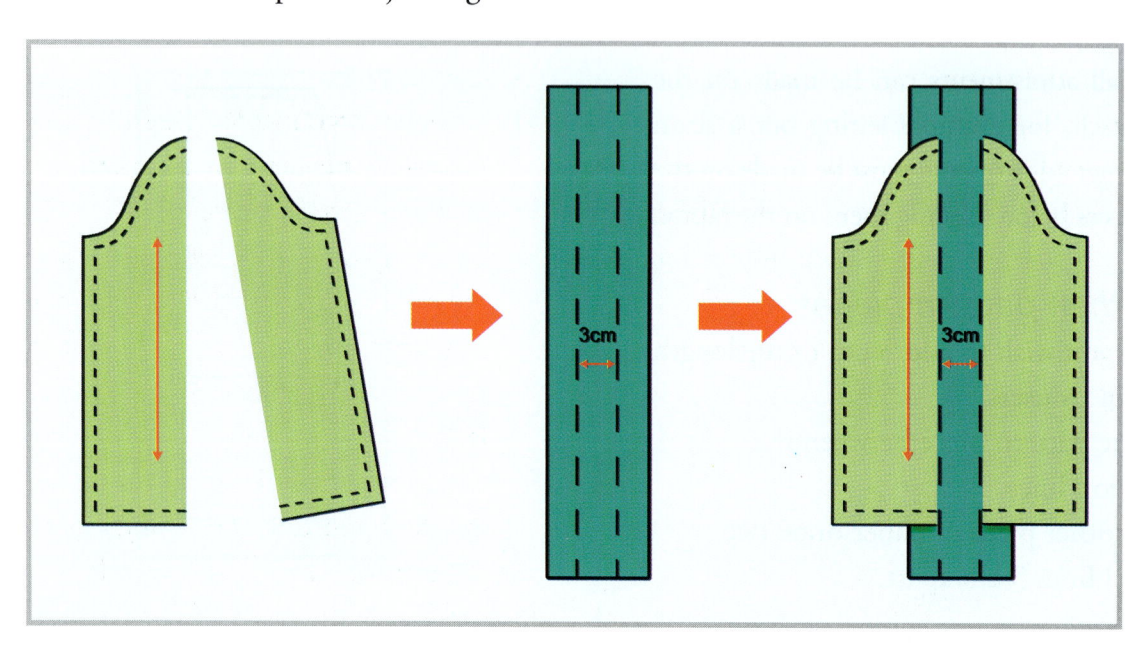

Increasing width

326

Using straight-of-grain lines

Lay the pattern pieces on the fabric so that the straight-of-grain arrow looks parallel with the selvedge edge. Pin one end of the arrow to the fabric. Measure from each end of the straight-of-grain arrow to the selvedge edge; when both ends are exactly the same distance from the selvedge pin the other arrow head in place. Then pin the rest of the pattern piece down.

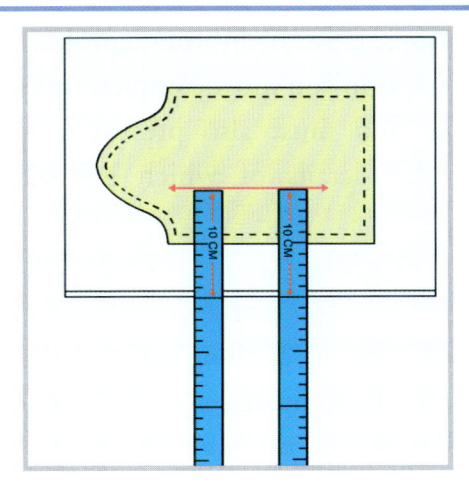

Laying pattern pieces using straight-of-grain lines

Cutting out
See Chapter 36, page 288.

Transfer of pattern markings
It is vital to transfer important pattern markings from the tissue pattern pieces to the fabric. There are three main methods for doing this.
1. Tailor tacks should always be used to show the location of important points on the pattern, e.g. end-of-bust darts, position of button and buttonholes etc.
2. Waxed paper and a tracing wheel are generally used to mark the sewing line. Two sheets of waxed paper are used, one against each piece of fabric on the wrong side. The wheel is rolled along the sewing line and the line comes out on the fabric and can be used as a guide to keep tacking and sewing straight.
3. Tailor's chalk is sometimes used for marking the sewing line when pattern pieces have no seam allowance included.

Fitting
It is important that the garment you are making fits well before you sew it up. A garment that fits well will:
- be neither too loose nor too tight
- hang well
- be comfortable and look well

Guidelines
- tack the large pieces, i.e. front and back pieces, of the garment together
- tack any darts in place
- try the garment on: sit, bend, raise your arms to see that the garment is not too tight; you will be able to loosen or tighten the garment by letting out or taking in some of your seam allowances
- check that shoulder seams lie on the shoulder and that the armhole is not too tight or too loose
- check that the waist of the garment is on the natural waistline

- mark alterations with chalk or pins; remove the garment and tack alterations; try garment on again before sewing

Pressing

Pressing is important to give your garment a finished, crisp look. Pressing is different from ironing: with pressing you lift the iron and press down on the fabric, lift and press etc.

Guidelines

- press after each stage of construction, e.g. after sewing the side seams; do not wait until the end to press everything as you won't get as good a result
- usually press on the wrong side; if you must press on the right side use a pressing cloth to protect the fabric
- use a steam iron set at the correct temperature for the fabric
- give the garment a final overall press when it is fully complete

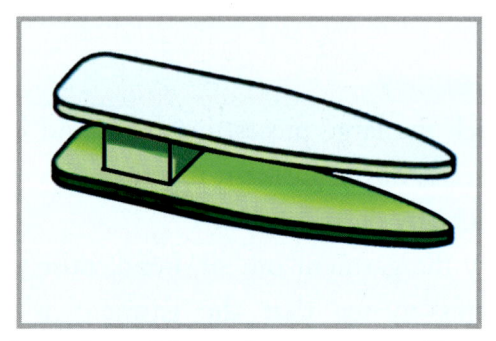

A sleeve board is used to press small areas of garments such as collars and sleeves.

TEXTILE SKILLS PROJECT

While there will be questions on Textile studies (Unit 5) on the final examination paper, the textiles skills *option*, like the other two options, is assessed through project work. For this option you must complete an item of clothing. You must also present a written folder, the details of which are outlined below.

Brief

Make an item of clothing which includes a minimum of two processes such as application of a collar, sleeves, zip, buttonholes etc.

Support folder

In your support folder you should include the following information:

- List the reasons for your choice of garment (e.g. fashion, shape suits your figure, for a special occasion/use etc.)
- Give the details of the pattern you used and any changes you made to it
- Give the details of the fabric you used (e.g. cost, amount you needed, type of fabric); draw a wash care label for your garment
- Evaluate your item of clothing under the following headings:
  How well is it finished? Evaluate every part of the garment: seams, seam finishes, collar, zip, buttons, buttonholes, hems, pockets etc.
  How well does it fit? Are there any puckers? Is the garment too tight or loose?
  What would you change next time round?
- You could complete a fabric burning test to identify your fabric: present your results
- What sewing equipment did you use? You could draw diagrams of some of the more unusual equipment, e.g. sleeve board
- Give details of the notions you had to purchase to complete the garment: list and price them (e.g. buttons, zip etc.)

THE TEXTILES INDUSTRY

The main textile industries in Ireland are:
- Irish tweed, e.g. Donegal tweed
- Irish linen, e.g. Belfast linen
- knitwear, e.g. Aran sweaters
- fashion designers, e.g. Paul Costello, Louise Kennedy, Quinn & Donnelly, John Rocha etc.
- Irish woollen carpets, e.g. Navan carpets

For those interested in fashion design or in the textile industry generally (weaving, printing etc.) a number of colleges around the country offer courses ranging from certificate to degree level. These include the National College of Art and Design, various post-leaving certificate colleges and FÁS training centres around the country. Ask your career guidance teacher for details.

Now test yourself at *www.my-etest.com*

picture credits

For permission to reproduce photographs, the author and publisher gratefully acknowledge the following:

Art Directors and TRIP: 5; 13 left; 29; 31; 32; 36 left; 44; 47; 69; 70; 75 both; 77 both; 81 all; 83 both; 84; 88; 89 both; 95; 96; 97 all; 98; 103; 104 both; 105 all; 113; 114; 115 top; 116 all; 117 all; 125 both; 128; 130 both; 132; 134; 135; 136; 138; 141; 144; 145; 146; 147; 150 both; 151; 163; 165; 167; 176; 180 left; 180 bottom right; 181 top left; 182 both; 184; 188 top right; 188 bottom right; 189 both; 191 left; 196 top; 198 all; 204; 205; 208; 213; 218; 224 both; 226 bottom left: 226 bottom right; 230; 232; 235; 240 all; 241; 242; 243 top left; 243 top right; 245 bottom; 247; 252 both; 253 both; 254 both; 255; 256 both; 266 left; 267 right; 268; 271; 272; 274 left; 276; 277; 283; 293 both; 295; 311; 319 top right; 319 bottom left: 321 all; 323; 329.

Advertising Archive: 180 top right; 181 top right; 181 bottom left; 181 bottom right.

Corbis: 108 © Michael Boys; 188 left © Carl and Ann Purcell; 209 left © Tom Stewart; 210 © Roy Morsch; 227 right © Rodney Hyett, Elizabeth Whiting & Associates; 273 left © Richard Hamilton Smith; 274 right © Geray Sweeney; 328 © Dann Tardif.

Digital Vision: 3; 27; 33; 36 right; 186; 209 right; 225 bottom; 229; 264 right.

ImageFile: 22 left © Mark Romine; 22 right © Jaume Gual; 23 left © James McLoughlin; 23 right © Stuart Pearce; 23 far right © Jim Toomey; 24 © Juan Manuel Silva; 25 © Esbin Anderson.

Marks and Spencer: 226 top left; 226 top right; 264 left; 266 right; 269; 275.

TopFoto.co.uk: 2 top right; 2 bottom right; 267 left; 273 right.

Other photos: Biophoto Associates/Science Photo Library: 10 top, 11, 13 right; British Dental Association: 197; Electrolux: 243 bottom left, 244, 245 top; Eyewire: 10 bottom, 196 bottom; Fired Earth: 225 top, 227 left, 243 bottom right; Getty Images/Nathan Bilow: 191 right; Glanbia Foods: 80 top, 82, 106; Greenpeace: 115 bottom; Mick Cullen: 223; PA Photos: 214 left; Paul Escudier: 312; Rex Features: 212, 214 right; Sheelin Irish Lace Museum: 319 top left; St. Mary's Hospital Medical School/Science, Photo Library: 2 bottom left; The Irish Image Collection: 223.

Picture Research: Image Select International Ltd.